Studies in History and Philosophy of Science I

Scientiarum Historia et Theoria Studia, volume 1

Roberto de Andrade Martins

Studies in History and Philosophy of Science I

Extrema
Quamcumque Editum
2021

ISBN: 978-65-996890-1-7

Summary

Foreword 1

The early history of dimensional analysis: I. Foncenex and the composition of forces 5

The early history of dimensional analysis: II. Legendre and the postulate of parallels..41

Experimental studies on mass and gravitation in the early twentieth century: the search for non-Newtonian effects................................77

The search for an influence of temperature on gravitation.............105

Philosophy in the physics laboratory: measurement theory versus operationalism 137

FOREWORD

This book contains a set of previously unpublished papers authored by me, Roberto de Andrade Martins. Some of those papers have been written several decades ago but they have not hitherto been published for a variety of reasons that will not be described here. Due to my advanced age and other current circumstances, it would not be wise to postpone the publication of those works even more.

(1) The early history of dimensional analysis: I. Foncenex and the composition of forces
(2) The early history of dimensional analysis: II. Legendre and the postulate of parallels

Those two papers were written in 1981. Their main results were published in summary form in the same year.[1] Another paper by the author, in Portuguese, added further information on the subject.[2] However, the detailed descriptions presented in the papers that are now brought to light have never been published before. In the interim, following the hints presented in the 1981 and 2004 published papers, other authors have mentioned Foncenex's work. However, the content of the articles was not revised to include more recent publications on the subject

---

[1] MARTINS, Roberto de Andrade. The origin of dimensional analysis. *Journal of the Franklin Institute*, **311**: 331-337, 1981.

[2] MARTINS, Roberto de Andrade. A busca da ciência a priori no final do século XVIII e a origem da análise dimensional. Pp. 391-402, in: MARTINS, Roberto de Andrade; MARTINS, Lilian Al-Chueyr Pereira; SILVA, Cibelle Celestino; FERREIRA, Juliana Mesquita Hidalgo (eds.). *Filosofia e História da Ciência no Cone Sul: 3º Encontro*. Campinas: AFHIC, 2004.

(3) Experimental studies on mass and gravitation in the early twentieth century: the search for non-Newtonian effects

Most of this paper was contained in a research project written in 1994, which was not published.[3] Its Introduction was added in 1995, when I presented a talk on this subject at the Laboratoire de Gravitation et Cosmologie Relativistes, Université Pierre et Marie Curie (Paris, France); the complete paper has not been previously published. When the paper was written, I had already published accounts of some episodes described in sections 2 and 4, in Portuguese.[4] Afterwards, I published some detailed papers addressing specific episodes shortly described here.[5] I have not

---

[3] Only a short version, in Portuguese, was published a few years later: MARTINS, Roberto de Andrade. Estudos experimentais sobre gravitação no início do século XX. Pp. 393-401, in: *VI Seminário Nacional de História da Ciência e da Tecnologia. Anais*. Rio de Janeiro: Sociedade Brasileira de História da Ciência, 1997.

[4] MARTINS, Roberto de Andrade. Pesquisas sobre a absorção da gravidade. Pp. 198-213, in: *Anais do I Seminário Brasileiro de História da Ciência e Tecnologia*. Rio de Janeiro: Museu de Astronomia e Ciências Afins, 1986; MARTINS, Roberto de Andrade. Os experimentos de Landolt sobre a conservação da massa. *Química Nova* **16** (5): 481-90, 1993.

[5] MARTINS, Roberto de Andrade. The search for gravitational absorption in the early 20th century. Pp. 3-44, in: GOEMMER, H., RENN, J., & RITTER, J. (eds.). *The expanding worlds of general relativity (Einstein Studies, vol. 7)*. Boston: Birkhäuser, 1999; MARTINS, Roberto de Andrade. Majorana's experiments on gravitational absorption. Pp. 219-238 in: EDWARDS, Matthews R. (ed.). *Pushing Gravity: New Perspectives on Le Sage's Theory of Gravitation*. Montreal: Apeiron, 2002; Russian translation: MARTINS, Roberto de Andrade. Opyty Majorany po poglosceniju gravitacii. Pp. 76-99, in: IVANOVA, M. A. & SAVROVA, P. A. (eds.). *Poiski mekhanizma gravitacii*. Nizhnij Novgorod: Izd. Ju. A. Nikolaev, 2004; MARTINS, Roberto de Andrade. Gravitational absorption according to the hypotheses of Le Sage and Majorana. Pp. 239-258, in: EDWARDS, Matthews R. (ed.). *Pushing Gravity: New*

expanded or updated the content of the paper, except for the addition of references to my own later publications on the subject.

(4) The search for an influence of temperature on gravitation

This article is a detailed exposition of one of the episodes dealt with in the previous paper. This work was written during the years 1995-1996, while the author was a visiting scholar of the Department of History and Philosophy of Science, University of Cambridge; and a visiting fellow of Wolfson College. It is now published for the first time. The content of the paper has not been updated or complemented; only slight changes were made.

(5) Philosophy in the physics laboratory: measurement theory versus operationalism

This paper was presented at the Eighth International History, Philosophy, Sociology & Science Teaching Conference, University of Leeds, England, July 15 to 18, 2005. It is published here for the first time. The author had previously published two papers that addressed some facets of the subject.[6] The content of the paper has not been updated or complemented; only slight changes were made.

Perspectives on Le Sage's Theory of Gravitation. Montreal: Apeiron, 2002, MARTINS, Roberto de Andrade. Searching for the ether: Leopold Courvoisier's attempts to measure the absolute velocity of the solar system. *Dio: The International Journal of Scientific History*, **17**: 3-33, 2011; MARTINS, Roberto de Andrade. Émile Meyerson and mass conservation in chemical reactions: *a priori* expectations versus experimental tests. *Foundations of Chemistry*, **21** (1): 109-124, 2019.

[6] MARTINS, Roberto de Andrade. Os elementos apriorísticos dos processos de medida. *Revista de Ensino de Física* **6** (2): 35-51, 1984; MARTINS, Roberto de Andrade. Measurement and the mathematical role of scientific magnitudes. *Manuscrito* **7** (2): 71-84, 1984.

About the author

Roberto de Andrade Martins is a Brazilian scholar. His research subjects are: foundations of physics (especially relativity theory), history of science (especially history of physics, mathematics, chemistry, astronomy and biology), and philosophy of science. He has published over 200 works on those subjects, and he has supervised over 30 dissertations and theses on his research subjects.

He obtained his first degree in Physics at the University of São Paulo (USP, 1972), and a PhD in Logic and Philosophy of Science at the State University of Campinas (UNICAMP, 1987). He was a member of the faculties of the State University of Londrina (UEL), of the Federal University of Paraná (UFPR) and of the State University of Campinas (UNICAMP), where he worked throughout most of his academic career. After retiring from UNICAMP, in 2010, he was a visiting professor of the State University of Paraíba (UEPB), of the University of São Paulo at São Carlos (USP) and of the Federal University of São Carlos (UFSCar). Since 2016 he is a collaborator of the Federal University of São Paulo (UNIFESP).

He was a research fellow of the Brazilian National Council for Scientific and Technological Development (CNPq) for more than 40 years. He was president of the Brazilian Society for History of Science (SBHC) and of the South Cone Association for Philosophy and History of Science (AFHIC).

THE EARLY HISTORY OF DIMENSIONAL ANALYSIS:
I. FONCENEX AND THE COMPOSITION OF FORCES

Roberto de Andrade Martins

Abstract: This paper describes a forgotten episode of the history of dimensional analysis. An article published in 1761 by François Daviet de Foncenex contains the first known attempt to derive a physical law – the parallelogram rule of forces – using the principle of homogeneity. The motivation of that work was the wish to provide an *a priori* proof of the basic laws of mechanics. The context and consequences of the paper are described. It is shown that this attempt was not grounded upon clear and solid assumptions and that its basic ideas were implicitly in conflict with the conceptions of that time.
Keywords: dimensional analysis; parallelogram of forces; laws of mechanics; a priori proof; Foncenex, François Daviet de

1. INTRODUCTION

Dimensional analysis is a method designed to produce scientific quantitative laws from a formal condition: the requirement of dimensional homogeneity (see Langhaar, 1951; Sédov, 1977). This technique received much attention in the last quarter of the 19th and in the first half of the 20th century. Today, it is not a very popular subject, although it has shown a moderate success in the realm of highly complex phenomena,

MARTINS, Roberto de Andrade. *Studies in History and Philosophy of Science I*. Extrema: Quamcumque Editum, 2021.

where a detailed theory of the processes is missing – such as in fluid dynamics and stellar structure.

Some books on dimensional analysis contain information about the history of this subject, and a few authors have sketched several components of its evolution (Ravetz, 1961; Macagno, 1971; Higgins, 1957). Although too little has hitherto been written on the development of dimensional analysis, the following outlook of its history emerges. The study of the dimensions of physical magnitudes has undergone a series of metamorphoses, and was linked to several subjects, such as: (i) a concern about unit conversion, and the use of dimensions to test the homogeneity of formulas (Fourier); (ii) the theory of models and similitude relations, studied by Galileo and Newton, and later linked to the method of dimensions, in the 19th century (Bertrand, Ledieu); (iii) the application of the method of dimensions, in the second half of the 19th century, to the derivation of formulas for complex phenomena (Rayleigh, Reynolds), and later to the foundations of physics (Rayleigh, Jeans, Einstein) and applied physics (Buckingham, Riabouchinsky) at the beginning of the 20th century; (iv) parallel to the development of the technique of dimensional analysis, the study of "absolute" measurement (Ampère, Weber, Gauss) led to the study of the dimensions of electromagnetic magnitudes (Maxwell, Jenkins, Vaschy, Hertz); the discussions about this subject soon became intermingled with controversies about ether models and the "essence" of electricity and magnetism. From those controversies came the popularity and stimulus for the study of dimensional concepts in the late 19th century; most textbooks on electromagnetism of that period included a section on the theory of physical dimensions, and this did not occur in other fields of physics; (v) in the two last decades of the 19th century, some authors tried to establish the foundations and to systematize the study of physical dimensions (Herwig, Vaschy, Piochon); (vi) in the first decade of the 20th century, interest on the ether and discussions about electromagnetism declined, and ultimately dropped to the background; after 1910, most authors ignored the previous history of dimensional analysis. With the publication of Bridgman's book (Bridgman, 1922), the modern period of dimensional analysis begins, and previous works were gradually forgotten.

This seems to be a correct, although incomplete, sketch of the history of the study of physical dimensions. Some interesting episodes have been left outside of the account, and the names of many contributors to this field have been omitted.

But even the above sequence of episodes never received a detailed historical study. Why has the history of this subject been hitherto so neglected? Two main reasons seem to have led to this state of affairs: (i) Most physicists today scarcely know that dimensional analysis exists, and most scientists probably think that this is a dead subject; its importance in current scientific research is negligible, and accordingly not many people would be driven to search its history; (ii) To most modern scientists – and even historians – it also seems that this subject has never been an important one, and that no deep problems have ever arisen concerning it; so, it does not deserve a detailed study. This is a wrong opinion, however.

Physicists will certainly be surprised when told that several famous scientists have devoted some of their time to this subject, studying its foundations, using dimensional analysis in fundamental research, and engaging into bitter discussions about physical dimensions. Besides, there are very deep aspects of dimensional analysis that have been overlooked and that do still allow fruitful foundational research – it is a forgotten but far from dead subject. For all those reasons, it seems that dimensional analysis still deserves a detailed study – and the best beginning seems to dig up its early history.

The specific aim of this paper is to study how dimensional analysis began. There is little doubt that the concept of physical dimensions now in use has been explicitly formulated for the first time by Fourier (1822, pp. 135-140). However, the use of dimensional analysis – the use of dimensional arguments in the derivation of equations – is not found in Fourier's works. This circumstance has led authors to think that dimensional analysis was created *after* Fourier's work. It has even occurred that someone who has made extensive use of dimensional analysis was credited with its origin. Thus, Riabouchinsky (1911) pointed to Rayleigh's 1899 work on capilarity as the first instance of this technique. This is far from true.

Other authors have tried to find the roots of dimensional analysis much time *before* Fourier, in the works of Galileo and Newton (Higgins, 1957, p. 331; Larmor, 1926, pp. 736-738; Ravetz, 1961, p. 9). It is true that Galileo has contributed to the theory of models – a field later linked to dimensional analysis – and that Newton has demonstrated a principle of mechanical similitude that was afterwards appplied to the theory of models and associated to dimensional reasoning. But both the concept of dimensions of physical magnitudes, and their use together with the principle of homogeneity to derive scientific laws, have arisen much later than that.

The use of dimensional analysis presupposes the use of functions and mathematical analysis within science. This did not occur in physics before the 18th century. Hence, it would not be wise to search for any instance of dimensional analysis before that century.

It seems natural to think that dimensional analysis must have been created after Fourier's elaboration of the concept of physical dimensions, in the 19th century; but actually the *use* of this technique has preceded the formulation of a *theory* of dimensions. The two earlier instances of the use of dimensional analysis that I have been able to find have appeared in an article on the foundations of mechanics, signed by François Daviet de Foncenex, published in the scientific proceedings of the Turin Academy (Foncenex, 1761); and in the *Elements of geometry* of Adrien Marie Legendre (1794). It seems to me that those two contributions to dimensional analysis have not hitherto been noticed by historians,[1] although both of them have been known to historians of other fields,[2] and they were sometimes cited by 19th century authors.[3]

Those two episodes, together with their historical context and consequences, will be studied in this paper and in its follow-up.[4] It will be shown that the origin of dimensional analysis was the search for *a priori* proofs of the fundamental laws of mechanics and geometry – specifically, the law of composition of forces, and the postulate of parallels.

In order to understand the significance of the article ascribed to Foncenex (hereafter called 'the Turin paper', for reasons that will become clear later), it is necessary to state the situation of mechanics – and in particular of the law of addition of forces – in the 18th century.

---

[1] This paper was written in 1981. Its main results were published in summary form in the same year (Martins, 1981). Another paper by the author, in Portuguese, added further information on the subject (Martins, 2004). However, the detailed descriptions presented in the current paper and in the next one of this volume have not been published up to now, that is, forty years later. In the interim, following the hint presented in the 1981 published paper, other authors have mentioned Foncenex's work. I will not present here a review of the studies on this subject published after 1981. Except for this note, the present chapter only reproduces the content of the original 1981 manuscript.

[2] The relevance of those works for the history of non-Euclidian geometry and mechanics has been remarked by Roberto Bonola (1955, pp. 53, 55-60, 197-199). However, this author has not noticed that Foncenex and Legendre have provided the first instances of dimensional analysis.

[3] Reference to the relation of Legendre's work and dimensional analysis may be found in Ledieu (1883) and Pionchon (1891, pp. 228-234).

[4] Roberto de Andrade Martins. The early history of dimensional analysis: II. Legendre and the postulate of parallels, in this volume.

2. THE PARALLELOGRAM OF FORCES

Newton's mechanics was ostensibly grounded upon three "axioms of motion" (Newton's three laws). The law of force composition (the parallelogram rule) is presented in Newton's *Principia* as a corollary to his laws of motion (Newton, 1952, p. 15), not as an independent law. However, it *is* an independent law, and cannot be derived from the three laws of motion.

The composition of uniform motions had already been studied by Galileo, and the idea can be traced back to Greek authors – indeed, it appears in the pseudo-Aristotelian *Mechanics*. Parallelograms of force have appeared several times in the 16th and 17th centuries, but it was only in 1687 that this law has been justified and applied to mechanical problems. It was simmultaneously presented by Newton, Varignon, and Lamy (Crowe, 1967, pp. 2, 13-14; Costabel, 1966; Montucla, 1802, pp. 609-610).

Once stated and applied with success, the rule was accepted, but its geometrical dress bothered many authors. Could this law be proved by a geometrical argument? Was it possible to provide an *a priori* proof of this law ? The usual answer was: Yes. Many attempts to devise a simple and correct proof of the law have arisen, from Newton's time to the 20th century (Dugas, 1950). One of the most famous was the one presented by Daniel Bernoulli (1726). As will be seen below, it has indirectly influenced the composition of the Turin paper.

In the 18th century, the word 'physics' was related to the idea of empirical studies; mechanics was not regarded as part of physics, by French researchers – it was considered to be a part of mathematics. At that time, all the branches of mathematics were believed to contain correct *a priori* knowledge. It was not thought to be a formal and conventional science, as we now believe. This change of outlook came only after the rise of non-euclidian geometry. People thought that mathematics should be grounded upon a few intuitive or apoditic axioms, and everything should be derived from those principles and definitions, by rigorous proof. It was expected that some non-evident principles of mathematics – such as the fifth postulate of Euclid's geometry – would be either eliminated as unnecessary, or derived from other clearer *a priori* truths. Mechanics, as a part of mathematics, should follow the same model.

This attitude can be clearly noticed in d'Alembert's work – and, as will be seen below, those ideas have strongly influenced the Turin paper. D'Alembert's attitude to mechanics follows his general ideas about science (d'Alembert, 1805, p. 30). He believed that science should be grounded on true principles – simple and recognized facts that can neither be denied nor explained – such as the impenetrability of bodies, in mechanics, which he thought to be the source of their mutual actions. Everything else should be derived from those simple principles.

In the preface of his *Traité de dynamique*, d'Alembert stated his opinion:

> The safety of the [parts of] Mathematics is an advantage that those sciences borrow from the simplicity of their subjects. It is necessary to admit that, since not every part of Mathematics has an equally simple subject, likewise the appropriate certitude, that which is founded upon Principles necessarily true and evidents by themselves, does not belong equally and in the same way to all those parts. Many of them, grounded on Physical Principles, that is, upon Experimental truths, or upon mere hypotheses, are just what could be called, let us say, of an Experimental certitude, or are even mere suppositions. To say more exactly, we can only regard as marked by the stamp of evidence those that deal with the calculus of magnitudes and the general properties ofextension, that is, Algebra, Geometry, and Mechanics. (d'Alembert, 1743, p. i)

D'Alembert stated that the foundations of Mechanics have been neglected: its principles are either obscure in themselves, or obscure demonstrations are provided for those principles (d'Alembert, 1743, p. iv). He proposed to reduce the number of the principles, to deduce them from clearer notions, and to apply them. He uses in his mechanics three fundamental principles: the law of inertia, the principle of composition of motions, and the equilibrium law (he used this name to refer to the collisions of bodies, not to the lever).

In the second edition of his *Traité*, d'Alembert discussed the problem raised by the Academy of Berlin: 'Are the laws of motion and equilibrium of bodies necessary or contingent laws?' (d'Alembert, 1758, pp. xxiv-xxix). He divided the problem in two parts: (1) Which are the laws of motion and equilibrium that would follow necessarily from the most basic properties of matter and motion? (2) Are those the observed laws? According to d'Alembert, it could happen that God would choose to apply to the world not the simplest laws, but other different laws, and this is the motivation of the second question. After developing his arguments, he reaches the final answer:

> From all those reflections, it follows that the known laws of statics and mechanics are those that result from the existence of matter and motion. But experience proves that those laws are indeed observed in the bodies around us. Therefore the laws of equilibrium and motion, such as those that observation inform us, are necessary truths. (d'Alembert, 1758, p. 397)

It seems to me that this is clearly a Cartesian attitude,[5] although sometimes d'Alembert criticizes Descartes. See for instance this ironical sentence: "... but nobody ignores that the Cartesians (a sect that today has almost disappeared)..." (d'Alembert,1805, p. 369; d'Alembert, 1743, p. v).

Since d'Alembert believed that the basic laws of mechanics could be derived from other simpler ideas, the problem was reduced to finding those simple ideas and the most plain

---

[5] I agree with Hankins' opinion on this point (Hankins, 1970). See, however, Cane's criticism and Hankins' reply (Cane, 1976; Hankins, 1976).

derivations of the laws. D'Alembert chose the law of composition of motions as one of the basic principles of his mechanics; therefore, part of his job was to prove it from simple ideas. He provided one demonstration in his *Traité*; a simpler derivation was presented later (d'Alembert, 1743, p. 22; d'Alembert, 1761-1780, vol. 6, pp. 360-369).

3. THE TURIN PAPER: MOTIVATION AND CONTENT

D'Alembert's ideas have greatly influenced the Turin paper. The motivation of that article is exactly the same as d'Alembert's. At the beginning of the paper, its aim is stated: to prove the laws of inertia, of composition of forces, and of equilibrium; and to answer the question: are the laws of mechanics necessary or contingent truths? (Foncenex, 1761, p. 299).

The main source of the ideas in the Turin paper seem to be d'Alembert's works. In this paper, d'Alembert is cited several times, and called 'un très-grand Géomètre' and 'l'homme Célèbre' (Foncenex, 1761, p. 299); the *Traité de dynamique* is cited, and at another point the paper refers to d'Alembert's article on *force* in the *Encyclopédie*,[6] and calls him 'illustre Ecrivain' (Foncenex, 1761, p. 304). He again cites d'Alembert and refers to his *Opuscules mathematiques*, at the same paragraph where a reference to Bernoulli may be found (Foncenex, 1761, p. 313). Since Bernoulli is cited by d'Alembert in that work, it seems likely that the author of the Turin paper did not read Bernoulli's original demonstration of the parallelogram law. Besides Bernoulli and d'Alembert, the only author cited in the Turin paper is a certain Mr. Formey (probably Jean-Henri-Samuel Formey), whose ideas he criticized (Foncenex, 1761, p. 318).

While studying the question proposed by the Berlin Academy, the paper refers to d'Alembert's belief that God could

---

[6] I believe that he refers to the article on "Composition du movement" (d'Alembert, 1778-1779).

violate the rational laws of mechanics; but this opinion is not accepted. The following argument is presented: being part of Mathematics, Mechanics has laws as evident as those of Geometry, and those laws cannot be violated. God can choose to act upon the bodies, and to direct their motion. He could even choose to make all bodies to move in circles, without any apparent reason. However, in this case, God's action would be a new force, and this force would necessarily obey the laws of mechanics (Foncenex, 1761, pp. 299-301, 318-319).

The article is divided in five parts: the introduction, the first numbered section, 'On the Force of Law of inertia', the second 'On the composition of forces', the third 'On the principle of equilibrium', and the fourth 'On the Lever'. Although the lever law may be derived from other principles, in the last section the author chose to provide an independent demonstration of this law, because "it seems very difficult to decide whether we should make the equilibrium of the lever to depend on the composition of forces, or conversely to deduce the latter principle from the equilibrium of the lever" (Foncenex, 1761, p. 301). There are two relevant passages where the principle of dimensional homogeneity is used: the derivation of the law of force composition; and the lever law.

The proposed demonstration of the parallelogram rule begins with a *Lemma*: before proving the general law of composition of forces, the paper proposes a proof that two forces of equal intensity and applied to the same body have a resultant that is proportional to their intensity and to a function of the angle between them. We reproduce below this part of the article:

> Lemma. If two equal forces with their quantities and directions being represented by the lines *CA*, *CB*, act upon any body *C* [Fig. 1], it is evident that this body will not be able to obey at the same time to those two forces: because it cannot move at the same time along *CA*, and along *CB*; it will therefore take a direction *CM* different from *CA* and from *CB*, and the line *CM* must necessarily divide the angle *ACB* in two equal parts, because, the forces *CA*, *CB* being equal by

supposition, everything that disposes *CM* to approach *CA* will equally dispose it to approach *CB*. This having been set, it is also evident that we may imagine a third force *CM* that produces alone the same effect upon the body *C*, as *CA*, *CB* conjointly. Besides, the quantity [intensity] of the force *CM* cannot depend but on the quantity of *CA* or *CB* and of the value of the angle *ACB*, and consequently if we make $CA = CB = a$, $CM = z$, $ACB = \phi$, we shall have $z = funct. (a, \phi)$.

But the force *CM* being of the same nature as the [*force*] *CA*, it is necessary that they contain one same number of dimensions; that gives $z = CM = funct. (a, \phi) = a.funct.\phi$, since the dimension of ϕ is null. (Foncenex, 1761, pp. 305-306)

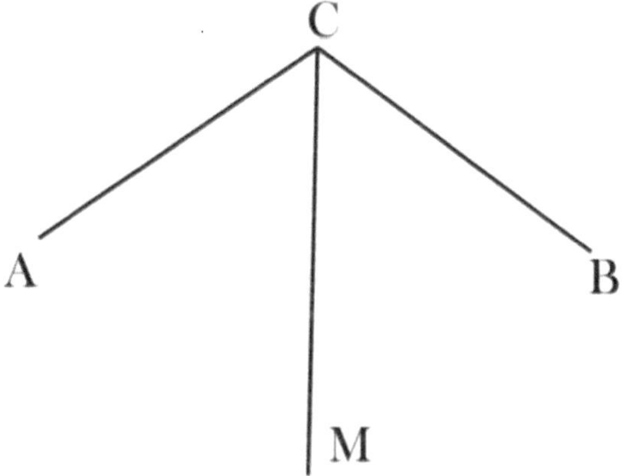

Fig. 1 – Composition of forces (Foncenex, 1761, planche 4, f. 1)

In a footnote the author remarked:

It follows from this that, the angle ϕ remaining constant, z is always proportional to a; we could in the same way demonstrate by this method, in a direct and very natural way, many theorems about the proportionality of the sides of figures, and a great number of other propositions of Geometry, and of Mechanics. (Foncenex, 1761, p. 306)

From the style of this note and other sentences of the paper, one may infer that the author was not aware of any previous use of this method. He says that "we *could* in the same way demonstrate... many theorems... of Geometry and of Mechanics" ("on pourroit de même démontrer par cette méthode..."), and this implies that this has not yet been done. At another place of the paper, we read: "The completely analytical demonstration that I have found has seemed to me otherwise worth finding its place here for its singularity" (Foncenex, 1761, pp. 301, 313). It seems therefore that the author did believe that his method was new.

A similar dimensional argument is used again twice in the paper. At one place, a variation of the demonstration of the parallelogram law is presented, where the author shows that the same results are reached if instead of forces proper, we consider the composition of what we now call moments; he first derives the basic lemma taking *mass* as the variable parameter, and afterwards does the same using *velocity* as the relevant parameter (Foncenex, 1761, p. 304). At another place, the author uses the same method to study the equilibrium of the lever (see Fig. 2), and starts by a new lemma:

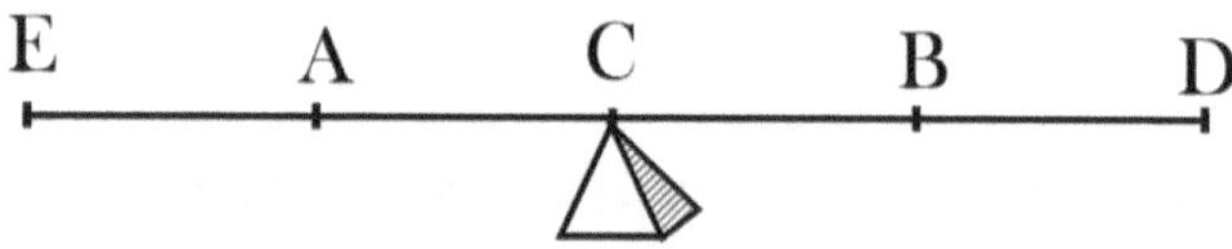

Fig. 2 – Lever equilibrium (Foncenex, 1761, planche 4, f. 7)

Lemma. If two equal forces $= p$ (as, for instance, two equal weights) act in parallel directions upon the lever AB at the points A and B at equal distances from the fixed point C, it is at once evident that the lever will be in equilibrium relative to the point C, since everything is equal on one side, and the other: I also say that the point C will support the same effort, as if the forces $p + p$ were directly applied to C; because this effort, or the force that would equilibrate them if it would act

> at C in the opposite direction, cannot depend but on the quantity p, and, if we want, of the distance CA, which I call x; this force will therefore be expressed by *fonct.* (p, x), and this we may demonstrate to be equal to *p.fonct.* x, as in the lemma of the Article I. (Foncenex, 1761, pp. 319-320)

It is very important to remark that in the case of the composition of forces (or momenta), one of the variables was the angle, with null dimension; but here, we have *two variable parameters which do not have null dimension.*

The dimensional arguments used in the Turin paper make use of some implicit assumptions. We may analyse the argument as if derived from the following premises:

TP (Turin Paper) 1 – Whenever two magnitudes are of the same nature, they have the same number of dimensions.

TP2 – Angles have null dimensions.

TP3 – Forces, masses and velocities have a number of dimensions different from zero.

TP4 – The dimensions of force are different from the dimensions of length.

TP5 – If z is a function of a and b; if z and a have the same number of dimensions (different from zero); and if b is a magnitude of null dimensions or has dimensions different from a and z; then we must have $z = a.f(b)$; that is, z must be directly proportional to a and to a function of b.

Those assumptions are sufficient – and, it seems to me, they are the most natural ones – to justify the steps where the Turin paper uses dimensional arguments.

But what exactly were the concepts of 'dimension' and 'number of dimensions' used here? Where did these ideas come from? There are two alternatives: either the author has created and used a concept of his own; or he is applying ideas previously developed and known. But when an author creates and uses for the first time a new concept, he usually elucidates its meaning. This elucidation would be particularly necessary in the present case, since there was a previous use of the word 'dimension' in geometry, and he should establish a distinction between his concept and the geometrical concept, if there was any difference between them. But in no place of the Turin paper can we find an elucidation of the concept of dimension; and in no other work signed by the same author can we find that elucidation. We may infer that the author is probably using a previously known concept, not a new one. Let us therefore recall what was the meaning of 'dimension' at that time.

4. THE CONCEPT OF DIMENSION IN THE 18TH CENTURY

In Diderot's *Encyclopédie* we find an article on 'Dimension' which was probably written by d'Alembert, who was the author of most scientific articles (d'Alembert, 1778-1779, vol. 10, pp. 1058-1059). It provides an obscure definition of dimension as "the extension of a body considered as measurable or susceptible of measure", and provides as instances: length, breadth, depth. In the same article we find the use of the word 'dimension' to denote algebraic powers or exponents (see also Rosenfeld & Cernova, 1967). The two uses of the word are related through a geometric instance: if a and b are two lines, then their product ab may represent the area of a rectangle with sides a and b; but a rectangle is a geometric figure with two dimensions, and lines have one dimension; hence, any product of two lines, or the second power of a line (or one dimension) may be regarded as corresponding to a figure of two dimensions. Also, the product abc may be interpreted as the volume of a solid that has three dimensions; as a general rule, the exponent or number of linear factors in a geometrical formula will correspond to the number of dimensions of the geometrical entity related to that formula, and this establishes a relation between algebrais powers and geometric dimensions. From this use have arisen in Antiquity our expressions 'square' for the second power, and 'cube' for the third power of any quantity.

Within geometry (but only in this field) the first assumption of the Turin paper (TP1) was well known and used. Each kind of geometric entity was supposed to have a specific number of dimensions: solids have three dimensions, surfaces have two, lines have one, points have none. There are no other geometric dimensions. Since only homogeneous quantities can be compared to one another, added together or divided by one another,[7] then, in geometry, the necessary and sufficient condition for the possibility of comparing, adding and dividing two quantities was their equality of number of dimensions. Hence, the principle of homogeneity became a condition about the number of dimensions of the concerned quantities. This requirement of dimensional homogeneity was widely used in analytic geometry, since the time of Fermat and Descartes (Fermat, 1891-1896, vol. 1, pp. 91-103; Descartes, 1664, pp. 67, 77, 80).

The second assumption was the statement that angles have null dimensions. The geometrical status of angles was not altogether clear in Antiquity, and it remains obscure (Heath, 1956, vol. 1, 176-180). Angles are geometrical entities, no doubt. If they have null dimensions, then they share the same nature of0 points, and this does not seem acceptable. But if an angle is thought as the space included between two lines, it is of the same nature as a surface, and would have two dimensions; but since the area corresponding to an angle would always be infinite, angles do not have a finite ratio to any limited surface, and therefore angles and finite surfaces are not homogeneous

---

[7] The ancient ideas about magnitudes can be found in Euclid's *Elements* (Heath, 1956, vol. 2, 112-120).

quantities.[8] This is not an elementary problem, and we cannot try to solve it here. We just need to remark that in the 18th century it was generally accepted that angles had null dimensions, being similar to abstract numbers. Hence, this supposition, although problematic, was not new.

The last assumption of the paper (TP5) may be considered a consequence of the requirement of dimensional homogeneity of formulas; but although this principle was accepted in geometry, I have been unable to find any previous use of this condition to *derive* the form of an equation. Previous authors have only used the principle to *verify* formulas. Hence, this assumption of the Turin paper seems original.

The two remaining assumptions (TP3 and TP4) are remarkable, and show a departure from the purely *geometrical* concept of dimension. The concept of dimension is now applied to *physical* parameters. This is a bold step, and the author of the paper probably had no clear idea about this use. The number of dimensions was known and understood for geometrical entities and abstract numbers; but what could be the number of dimensions of *force*, or *mass* ? If they did have a definite number of dimensions, they would be dimensionally equal – and therefore homogeneous to – some kind of geometrical entity, and could therefore be equated or added to it in mathematical formulas. But this would be an unacceptable idea, at that time.

Notice that, at the time of publication of the Turin paper, the concept of dimension was quantitative (numerical) and not qualitative. Different *kinds* of dimensions were not discussed. 'Dimension' was exactly equivalent to 'geometrical dimension'. It is true that d'Alembert refers to the possibility of considering *time* as a fourth dimension,[9] but he states this as a mere curiosity,

---

[8] Homogeneous quantities must have a finite ratio. A line is not homogeneous to a surface because there is not a finite ratio between them (Heath, 1956, vol. 2, 112-120).

[9] In the *Encyclopédie* article referred above, d'Alembert stated: "I have said above that it is impossible to conceive more than three dimensions. A friend of mine believes that we can regard duration as

and this was certainly not a common idea. Our modern conception of different kinds of dimensions did not exist at that time.[10]

We may find some previous use of the concept of dimension within physics. One of them, indicated by Macagno (1971), appears in Descartes' speculations. But Descartes' use is obscure and had no influence on the later development of the concept of physical dimensions (Martins, 1981). Even if the author of the Turin paper knew Descartes' ideas, they could not have aided him.

Another instance, also remarked by Macagno, is that made by Euler, who explicitly talks about dimensions, and homogeneity conditions in mechanics (Euler, 1948). However, before studying his contribution, let us go back to earlier ideas.

In Antiquity, it was accepted that to multiply or to divide two heterogeneous magnitudes was absurd, except in the case of geometrical magnitudes (Bochner, 1963). So, while we ordinarily represent the lever law as the equality between two *products* of length versus force

$$F_1.L_1 = F_2.L_2 ,$$

Archimedes could only understand this law as an equality of *ratios* of homogeneous quantities:

$$F_1/F_2 = L_2/L_1 .$$

Although in the 13th and 14th centuries some authors did already define speed as the ratio of two concrete quantities – as space divided by time (Crombie, 1961) – Galileo in the 17th century still represented the relation between space, time and speed in a way that, in modern notation, can be rendered as:

---

a fourth dimension, and that the product of time versus a solid would somehow be a product of four dimensions; this idea may be contested, but it has some merit, it seems to me – at least that of novelty" (d'Alembert, 1778-1779, vol. 10, pp. 1058-1059).

[10] The idea only appeared with Fourier (1822, vol. 1, pp. 135-140).

$$d_1/d_2 = (v_1/v_2).(t_1/t_2),$$

for uniform motion (Galilei, 1842-1856, vol. 13, p. 152), and

$$e_1/e_2 = (t_1/t_2)^2$$

for uniformly accelerated motion (Galilei, 1842-1856, vol. 13, p. 168). He also presented the relation between mass, density and volume (Galilei,1842-1856, vol. 12, p. 21) in a way that we represent (in modern notation) as:

$$M_A/M_B = (d_A/d_B).(V_A/V_B).$$

This was the usual way of understanding physical relations: as equations between pure or abstract numbers, and not as relations between physical magnitudes of different kinds. That is also the notion which we may find in the works of d'Alembert, Carnot and Lagrange, shortly before or after the time of the Turin paper (d'Alembert, 1805, p. 404; Carnot, 1803, p. 11).

In the beginning of his book *Theoria motus corporum solidorum seu rigidorum*, Euler assumed the same idea. After defining velocity as the ratio of space divided by time (Euler, 1948, pp. 28-29), he asks: How could we divide space by time, since they are heterogenous quantities? We cannot say how many times is a time such as *ten minutes* contained in a space such as *ten feet* – and if we could divide a space by a time, then this time would be contained in that space. But Euler shows that all relations between speed, space and time may be reduced to ratios between homogeneous quantities and equations between pure numbers, in the same way that was envisaged by Galileo and other former authors; in this way, all difficulties disappear.

But if all physical laws are to be reduced to relations between numbers, no dimensional requirement can be applied to them, since there is no homogeneity restriction to the form of relations between adimensional quantities. For instance: according to our modern ideas, a physical equation such as

$$F = m.a$$

is dimensionally correct, but

$$F = m/a$$

would be wrong. But this second equation can be written as:

$$F_1/F_2 = (m_1/m_2).(a_2/a_1),$$

and this equations is correct, from the dimensional point of view. Hence, the Turin paper could not make use of an interpretation similar to Euler's.

However, at another place of his book, Euler used a different approach. He again discussed the problem of comparison of heterogeneous quantities, and also the problem of arbitrariness of units, proposing a method of *absolute measurement* of mechanical quantities (Euler, 1948, p. 82). As will be seen below, he attempted to reduce all physical magnitudes to lengths and abstract numbers.

Euler assumed, as we do, that weight and forces are homogeneous quantities; but he also stated that he would *use* weight as a measure of mass, because at each place they are proportional; and *he accordingly accepts them as homogeneous quantities* (Euler, 1948, p. 87). Since in mechanical equations there appear ratios of force to mass, those ratios become abstract numbers. Euler also assumes that times are always to be referred to (or divided by) the second, and hence, whenever a symbol t for time appears in an equation, an absolute number is to be understood by this letter.

Euler states that whenever forces appear in an equation, they are to be divided by the weight of the body to which they are applied, and hence only an absolute number will appear in the place of forces. Velocities are to be measured by the space traversed *in one second*. Euler then expresses velocities by spaces, and this amounts to regard velocities and lines as homogeneous quantities (Euler, 1948, p. 89). Hence, in all physical equations, only two kinds of quantities appear: either absolute numbers, or geometrical lines. Accelerations are also regarded as homogeneous to lines, since time is regarded as an abstract number (Euler, 1948, p. 90).

Euler remarked that it is easy to notice the homogeneity of the equations of motion, since the space traversed by a body, its velocity and acceleration are *linear quantities of one dimension* ("sint quantitates lineares et quasi unius dimensionis"), and times and the ratios of force per mass are considered as absolute numbers, which must be reckoned as of null dimension ("qui nullam dimensionem constituere sunt censendi") (Euler, 1948, p. 91).

Here we find, possibly for the first time, the application of the ideas of geometrical dimension and dimensional homogeneity to mechanical quantities and mechanical laws, although in a way completely different from Fourier's, for instance. Although there is no reference to Euler in the Turin paper, it is possible that its author was familiar at least with some of Euler's work, as will be seen below. But even if Euler's use of dimensions in mechanics were known to the author of the Turin paper, he could not be using those ideas in his derivation, since they are not compatible with assumptions TP3 and TP4. For Euler, as has been shown, forces are always to be divided by the weight of the body, and hence to be considered as quantities of null dimension; and if we accept this, no dimensional requirement can be applied to the equation of composition of forces, and nothing can be concluded from the relation $z = $ *funct.* (a,ϕ).

Notice also that in his derivation of the lever law, the author must assume that forces and lengths have different dimensions, and that it is not possible to produce a non-dimensional quantity from forces and distances; he is therefore assuming that forces are not geometrical entities, and that they do not have purely geometrical dimensions. Since, in that derivation, he again takes the force out of the function, he cannot assume that forces have null dimension. This is only compatible with the idea of different *kinds* of dimensionality – an idea that we do not find in Euler.

The ideas about the dimensions of physical magnitudes used in the Turin paper were therefore new and at variance with the conceptions of that time. But the author probably did not realize this, since he neither states that his ideas are new, nor elucidates his concept of the dimensions of physical magnitudes.

5. THE AUTHORSHIP OF THE TURIN PAPER

Let us now explain why up to this point we are mentioning "the Turin paper", instead of referring to Foncenex, its putative author. The reason is this: it seems that at least the *ideas* of that article are due to Lagrange. Let us see the relevant evidence.

Lagrange was born in Turin, in 1736, and at the age of 19 or 16 he began his teaching career in that same city, at an artillery academy (Delambre, 1867, vol. 1, p. ix). Foncenex was one of his students, there. At this time, Lagrange had already been influenced by Euler (Genocchi, 1883).

In 1757, at the age of 21, Lagrange joined Count Saluzzo di Menusiglio and Giuseppe Cigna to create the Academy of Sciences of Turin (Gorresio, 1883). In 1759 this Academy published its first volume of memoirs, called *Miscellanea philosophico-mathematica societatis privatae Taurinensis*. This volume contained an article on imaginary and complex numbers signed by François Daviet de Foncenex, with a note by Lagrange (Foncenex, 1759). This shows the close association betweem them at that time. That was Foncenex' first paper.

Lagrange sent this first volume to outstanding scientists and mathematicians, including d'Alembert. The later, in reply, sent to Lagrange the four first parts of his *Opuscules mathématiques* (d'Alembert, 1761-1780). In the first one, d'Alembert presented his demonstration of the parallelogram rule which was cited above. This may have been the stimulus for the composition of the Turin paper.

In the very first letter from d'Alembert to Lagrange, with the date of September 27, 1759, we find a reference to Foncenex (Lagrange, 1867-1892, vol. 13, pp. 3-4).

In 1759 and 1760, six new members joined the Turin Academy; among them we find Daviet de Foncenex (Gorresio, 1883). In 1761 the second volume of memoirs was published, now under the noble name of *Mélanges de philosophie et mathématique de la Societé Royale de Turin*. Here appeared the paper on the fundamental principles of mechanics *"par Monsieur le Chevalier Daviet de Foncenex"*.

It is possible that Lagrange has shown to Foncenex d'Alembert's work on the law of force composition, and that he discussed those ideas with him. It is also possible that he has suggested to Foncenex the main ideas of the paper, and stimulated his pupil to write the article, while he was himself busy with his researches on the theory of sound. This very bulky study was published at the same time as Foncenex's paper (Lagrange, 1760-1761).

Delambre states that, according to Lagrange himself, "he provided Foncenex with the analytical part of his Memoirs, leaving to him the care of developing the arguments about the formulas" (Delambre, 1867, p. xi). Genocchi states that Foncenex's paper on the principles of mechanics "is said to have been made by Lagrange or with his help" (Genocchi, 1883, p. 86; Genocchi, 1869). There must be some truth behind those rumours. It is remarkable that in his *Mécanique analytique* Lagrange refers to the Turin paper, but he does not cite Foncenex by name: "See the second volume of the *Mélanges de la Société de Turin*" (Lagrange, 1867-1892, vol. 11, p. 19). Let us also notice that other later authors do also refer to the paper without telling the author's name (Legendre, 1794; Fourier, 1888-1890, vol. 2, pp. 475-521; Laplace, 1878-1912, vol. 8, pp. 69-197). Was this because everyone knew that Foncenex was not the author? Maybe Lagrange did not mention Foncenex because the real author of the paper was himself. That is not conclusive evidence, however, since at this same place Lagrange *criticized* the demonstration of the principle of force composition presented in the Turin paper, and does not use it in his own book.

Let us add some information about Foncenex (Anonymous, 1857-1866; Anonymous, 1843-1847). François Daviet de Foncenex was born at Thonon, in 1734, being therefore two years older than Lagrange. His only relevant scientific papers were those cited above, published bu the Turin Academy. Shortly after the publication of the paper on the foundations of mechanics, and through Lagrange's influence, he was placed by the king of Sardaigne at the head of his Navy; in 1766, Lagrange told d'Alembert that Foncenex was at the sea. Afterwards he became governor of Sassai and Villefranche. In 1789, he published his third and last known scientific contribution: a description of an upwards thunderbolt rising from the Villefranche beacon. In 1792, he was accused of weakness or treason because he did not duly defend Nice, and he was arrested for one year. In 1799, the year of his death at Casals, an edition in book form of his work on the principles of mechanics was published in Turin. One of the biographical notes states that he left several manuscripts on algebra and geometry, but nothing is known about their content (Anonymous, 1857-1866).

6. THE INFLUENCE OF THE TURIN PAPER

The Turin paper did not produce any considerable immediate impact. Three papers, by d'Alembert, Laplace, and Fourier, have referred to it and corrected an analytical mistake of the article. Once the error is corrected, it is seen that the derivation of the paper does not conduce to the usual law of the lever (Bonola, 1955, pp. 181-199; Fourier, 1888-1890, vol. 2, pp. 475-521; Laplace, 1878-1912, vol. 8, pp. 69-197). It is quite interesting to notice that the law of force composition is now known to hold in any kind of geometry – since it is a differential law. But the lever law, which refers to an extended body, assumes different forms in different geometries. The corrected derivation of the lever law of the Turin paper is compatible with non-Euclidian geometries – a result later studied by Angelo Genocchi (1869a; 1869b; 1878). But it was only one century after the publication of the Turin paper that non-Euclidian mechanics was developed and studied by de Tilly, Mansion, Andrade, and others (see Bonola, 1955, pp. 181-199; Grigorian, 1960). Those developments, not directly linked to our subject, will not be described here.

Fourier's article is a proof that this author did know the Turin paper before he wrote his *Théorie analytique de la chaleur*, where he presented his considerations on dimentions of physical magnitudes. Therefore, it is possible that the Turin paper has influenced Fourier's later work on the dimensions of physical magnitudes and the homogeneity of physical laws.

The Turin paper was read by Legendre, and has motivated his work on dimensional analysis. This important development will be dealt in our next article.[11]

Within classical mechanics, the dimensional method used in the Turin paper did not become influential. The only direct effect that I have been able to find is the reproduction of the derivation of the composition of forces in the second edition of Poisson's *Traité de mécanique* (Poisson, 1833).[12] We shall present a detailed study of this case, since it illustrates how difficult it was to combine the assumptions of the Turin paper with the concepts of that period.

Contrasting with the Turin paper, Poisson takes the care of explicitly describing the principle of dimensional homogeneity, before using it. It is interesting to remark that Foncenex did not use the word 'homogeneity', and that Poisson uses it, while avoiding to refer to the 'dimension' of mechanical magnitudes. Let us quote Poisson:

> The equations that we shall consider will contain abstract numbers, such as the number π, logarithms, trigonometric lines, etc; they will also contain other quantities of several natures, that will also be represented by numbers expressing their ratios to arbitrarily chosen units, granted that each unit

---

[11] See the next paper of this volume.

[12] The first editon of Poisson's treatise (Poisson, 1811) does not contain a corresponding passage.

will be the same for every quantity of the same kind. Changing the magnitude of one or several units, the numbers that express the corresponding quantities will vary inversely as that magnitude, and, notwithstanding this completely arbitrary change, the equations that contain them must still hold. It is necessary, for this to happen, that their forms obey certain conditions, easy to verify in each particular case, and that are called, in the most general acception, the conditions of *homogeneity of the quantities*. Any equation that does not satisfy them, will be wrong for this reason, and must be rejected. (Poisson, 1833, vol. 1, p. 39)

As will become clear in the following quotation, Poisson regarded each unit as independent of the others, except in the case of the units of length, area, and volume. He did not try to reduce all magnitudes to a set of a few fundamental units, as Fourier did; this takes from his method all its practical value, and produces consequences that clash with modern dimensional analysis.

Thus, representing by F a given function, let us suppose that we have

F(f,f',...L,L',...m,m',...t,t',...) = 0;   (a)

$f, f',...$ being forces, $L, L',...$ lines, $m, m',...$ masses, $t, t',...$ times. If we represent by n, n', n'', n''' several abstract numbers, and if we reduce at the same time the unit of force in the ratio of one to n, the linear unit in the ratio or one to n', the unit of mass in the ratio of one to n'', the unit of time in the ratio of one to n''', the numbers $f, f',... L, L',... m, m',... t, t',...$ will become $nf, nf',... n'L, n'L',... n''m, n''m',... n'''t, n''' t',...$, and the equation (a) must still be valid, that is, one must still have

F(nf,nf',...n'L,n'L',...n''m,n''m',... n'''t, n'''t',...) = 0,

whatever may be n, n', n'', n'''. If the equation included surfaces $s, s',...$ and volumes $v, v',...$, their dimensions should be reported to the same unit as the lines $L, L',...$ and those quantities $s, s',... v, v',...$ would consequently become $n'^2s, n'^2s',... n'^3v, n'^3v', ...$ by the modification of this unit. (Poisson, 1833, vol. 1, pp. 39-40)

Here, Poisson explicitly referred to the relation between the units of length, area, and volume, and fails to mention other relations, such as that between velocity and length. Since he says that the equation must hold "whatever may be n, n', n'', n'''," this implies that the units of force, length, mass, and time, could be arbitrariyly and independently changed without affecting the equation. This does not correspond to our current notion.

Poisson presented an instance of his principle, showing that a particular equation formerly presented in his book did satisfy the homogeneity principle. However, the formula he tested only contained geometrical quantities. He next proposes a new rule:

> It is impossible that the equation (a) may contain a single quantity of some kind alone; when it contains two – for instance, two forces f and f' – and we solve [the equation] relative to one of them, obtaining
> f' = F(f,L,L',...m,m',...t,t',...),
> it is necessary, by the homogeneity of the quantities, that f be a factor of all the terms of the new function F, or, said otherwise, it is required that we have:
> f' = Nf;
> N being a factor that will contain no quantity of the nature of f and f', and will not vary with the unit of force. (Poisson, 1833, vol. 1, p. 41)

Notice that, if Poisson's principle of homogeneity was correct, then any formula such as

$$F = m.a = m.d^2x/dt^2$$

would be deemed wrong, since it contains only one quantity of each kind. Poisson uses equations such as this, in his book, but he did not discuss this problem. Actually, in the main text of his book, we may find one single use of his principle of homogeneity: in the derivation of the parallelogram rule. The demonstration follows the general lines of the Turin paper, and is probably derived from it, although Poisson did not refer to it.

Let us reproduce the relevant part of the argument, in order to compare it to the proof in the Turin paper.

> The resultant of two equal forces always cuts into two equal parts the angle comprised between their directions; because there would be no reason for it approaching more one of these two forces, or for its direction to leave their plane more to one side than the other; its direction is therefore known, and we need only to determine its magnitude.
>
> To find it, let MA and MB be the directions of the components, their common value being represented by P. Let also $2x$ be the angle AMB, and MD the direction of the resultant, in such a way that $AMD = BMD = x$. Its intensity cannot depend but on the quantities P and x; representing it by R, we shall have
>
> $R = f(P,x)$.
>
> In this equation, R and P are the only quantities whose numerical expression varies with the unit of force; according to the principle of homogeneity of quantities, it is therefore required that the function $f(P,x)$ takes the form $P\phi x$. Thus we have
>
> $R = P\phi x$;
>
> and the question is reduced to the determination of the form of the function ϕx. (Poisson, 1833, vol. 1, pp. 45-46)

Notice that Poisson did not use the assumption that angles have null dimension. Let us explicitly state his assumptions, in order to show how different they are from those of the Turin paper:

SDP1 – The units of each kind of mechanical quantity are arbitrary and independent of other units.

SDP2 – The equations of mechanics must remain valid if we multiply each kind of quantity appearing in them by arbitrarily chosen numbers (remarking, however, that the geometrical quantities are not independent of one another).

SDP3 – An equation cannot contain one single mechanical quantity of some kind; and when it contains only two, they will necessarily be proportional to one another.

In the specific instance of the derivation of the parallelogram law, Poisson arrives to the same result as the Turin paper, but their premises are completely different. Notice that the assumptions of the Turin paper are compatible with modern dimensional analysis, and those of Poisson are not. But at that time, Poisson's ideas were much more natural and acceptable than those of the Turin paper.

It seems that Poisson did not pay much attention to the consequences of his principles. It is likely that his only motivation was to provide a justification for the proof of the law of composition of forces.

7. CONCLUDING REMARKS

Since our main theme, dimensional analysis, was historically linked to the search for a proof of the law of composition of forces, let us briefly refer to the later phases of this subject.

In 1875 – a hundred and four years after the publication of the Turin paper – Darboux, while proposing a new demonstration of the law of force composition, presented a brief review of former works on the subject, and remarked: "Today, we seldom find a scientific journal where we do not find at least one demonstration of the parallelogram law" (Darboux, 1875; see also Aimé, 1836). Why did so many people attempt to find the proof of this law? Perhaps because it looked like a geometrical theorem (not like a physical law) and so one was tempted to derive it from *a priori* notions. Even in the 20th century 1941 we may find Birkhoff presenting a new derivation of this law, and emphasizing that its main ingredient is the *a priori* principle of sufficient reason (Birkhoff, 1941).

There was some difference between the attitudes of British and French authors regarding this subject.[13] While in France there was a wide acceptance of the possibility of providing *a priori* demonstrations of the laws of mechanics, British scientists usually regarded the laws of mechanics as empirical truths, and did not pay so much attention to those attempts of demonstration. In the early 19th century, Whewell presents a very modern and lucid discussion of the epistemological status of the laws of motion, showing that their general and abstract forms are indeed *a priori* truths; but that the particular formulations that render them applicable to the reality are empirical and *a posteriori* (Whewell, 1834). The specific relation between mechanics and geometry, and the conditions that allow us to produce an apparently geometrical derivation of the law of force composition are clearly and correctly discussed by Goodwin and de Morgan, some time later (Goodwin, 1847; Morgan, 1864). Let us remark that de Morgan uses a dimensional argument in his article.

The empirical approach to mechanics of British scientist, perhaps a legacy of Newton's misunderstood *hypotheses non fingo* (Bell, 1942), was not a fertile ground for the creation and development of methods that proposed to provide an *a priori* proof of the basic principles of mechanics. The French science of the 18th and early 19th centuries, however, deeply influenced by Descartes' rationalism, was probably the best field for the search of such methods – and this allowed the creation of dimensional analysis.

As we have seen, however, although the motivation for the creation of dimensional analysis was strong and clear, it did not have a good conceptual support, at the time of publication of the Turin paper. The attempt was premature and was not grounded

---

[13] One may consult some interesting British accounts of the differences between French and British science – and other cultural differences: Anonymous, 1820; Anonymous, 1821; Anonymous, 1821-1822.

on firm foundations. It made use of ideas in disaccord with those common at the time, and it was probably for that reason that the method was later wrongly stated by Poisson.

The most important and direct influence of the Turin paper was to stimulate Legendre's work on dimensional analysis; this subject will be dealt with in our following paper.

ACKNOWLEDGMENT

This work has been supported by the Brazilian National Council for Scientific and Technological Development (Conselho Nacional de Desenvolvimento Científico e Tecnológico – CNPq).

BIBLIOGRAPHIC REFERENCES

[ANONYMOUS]. State of science in England and France. *Edinburgh Review*, **34**: 383-422, 1820.

[ANONYMOUS]. English and French literature. *Edinburgh Review*, **35**: 158-190, 1821.

[ANONYMOUS]. Stewart's 'Introduction to the Encyclopaedia'. *Edinburgh Review*, **36**: 220-267, 1821-1822.

[ANONYMOUS]. Foncenet [*sic*], François Daviet de. Vol. 7, p. 321, in: MICHAUD, Joseph François (ed.). *Biographie universelle ancienne et moderne.* 21 vols. Bruxelles: Ode, 1843-1847.

[ANONYMOUS]. Foncenex, François Daviet de. Vol. 13, p. 240, in: HOEFER, Jean Chrétien Ferdinand (ed.), *Nouvelle biographie depuis les temps les plus reculés jusqu'a nos jours.* 46 vols. Paris: Firmin Didot, 1857-1866.

AIMÉ, M. Démonstration du parallélogramme des forces. *Journal de Mathématiques Pures et Appliquées*, **1**: 335-338, 1836.

BELL, Arthur E. Hypotheses non fingo. *Nature*, **149**: 238-240, 1942.

BERNOULLI, Daniel. Examen principiorum mechanicae, et demonstrationes geometricae de compisitione et resolutione virium. *Commentarii Academiae Scientiarum Petropolitanae*, **1**: 126-142, 1726.

BIRKHOFF, George David. The principle of sufficient reason. *Rice Institute Pamphlet*, **28**: 24-50, 1941; reprinted in vol. 3, pp. 778-804, in BIRKHOFF, George David. *Collected mathematical papers*. 3 vols. New York: American Mathematical Society, 1950.

BOCHNER, Salomon. The significance of some basic mathematical conceptions for physics. *Isis*, **54**: 179-205, 1963.

BONOLA, Roberto. *Non-Euclidean geometry: a critical and historical study of its development.* New York: Dover, 1955.

BRIDGMAN, Percy Williams. *Dimensional analysis*. New Haven: Yale University Press, 1922

CANE, Edric. Jean d'Alembert between Descartes and Newton. *Isis*, **67**: 274-276, 1976.

CARNOT, Lazare Nicolas Marguerite. *Principes fondamentaux de l'équilibre et du mouvement.* Paris: Bachelier, An XI [1803].

COSTABEL, Pierre. Varignon, Lamy et le parallélogramme des forces. *Archives Internationales d'Histoire des Sciences*, **19**: 103-124, 1966.

CROMBIE, Alistair Cameron. Quantification in medieval physics. *Isis*, **52**: 143-160, 1961.

CROWE, Michael J. *A history of vector analysis*. Notre Dame: University of Notre Dame Press, 1967.

D'ALEMBERT, Jean Baptiste le Rond. *Traité de dynamique.* Paris: David l'Aîné, 1743.

D'ALEMBERT, Jean Baptiste le Rond. *Traité de dynamique.* 2nd. edition. Paris: David, 1758.

D'ALEMBERT, Jean Baptiste le Rond. *Opuscules Mathématiques, ou mémoires sur différens sujets de*

Géométrie, de Méchanique, d'Optique, d'Astronomie &c. 8 vols. Paris: David, 1761-1780.

D'ALEMBERT, Jean Baptiste le Rond. Composition du movement. Vol. 8, pp. 763-765, in: DIDEROT, Denis; D'ALEMBERT, Jean Baptiste le Rond (eds.) *Encyclopédie ou dictionnaire raisonné des sciences, des arts et des métiers.* 3rd. edition. 39 vols. Genève: chez Jean-Léonard Pellet, 1778-1779.

D'ALEMBERT, Jean Baptiste le Rond. Dimension. Vol. 10, pp. 1058-1059, in: DIDEROT, Denis; D'ALEMBERT, Jean Baptiste le Rond (eds.) *Encyclopédie ou dictionnaire raisonné des sciences, des arts et des métiers.* 3rd. edition. 39 vols. Genève: chez Jean-Léonard Pellet, 1778-1779.

D'ALEMBERT, Jean Baptiste le Rond. *Oeuvres philosophiques, historiques et littéraires. 2. Essai sur les élémens de philosophie ou sur les principes des connoissances humaines.* Paris: J.-F. Bastien, 1805.

DARBOUX, Gaston. Sur la composition des forces en statique. *Bulletin des Sciences Mathématiques et Astronomiques*, **9**: 281-288, 1875.

DELAMBRE, Jean-Baptiste Joseph. Notice sur la vie et les oeuvres de M. le Comte Joseph-Louis Lagrange. Vol. 1, pp. viii-li, in: SERRET, Joseph-Alfred (ed.). *Oeuvres de Lagrange.* 14 vols. Paris: Gauthier-Villars, 1867-1892.

DESCARTES, René. *La géometrie.* Paris: Charles Angot, 1664.

DUGAS, René. *Histoire de la mécanique.* Neuchatel: Griffon, 1950.

EULER, Leonhard. *Theoria motus corporum solidorum seu rigidorum.* Bernae: Societatis Scientiarum Naturalium Helveticae, 1948.

FERMAT, Pierre de. Ad locos planos et solidos isagoge. Vol. 1, pp. 91-103, in: TANNERY, Paul; HENRY, Charles (eds.). *Oeuvres de Fermat.* 3 vols. Paris: Gauthier-Villars, 1891-1896.

FONCENEX, François Daviet de. Sur les quantités imaginaires. *Miscellanea Philosophico-Mathematica Societatis Privatae Taurinensis*, **1**: 113-146, 1759.

FONCENEX, François Daviet de. Sur les principes fondamentaux de la mechanique. *Mélanges de Philosophie et Matématique de la Société Royale de Turin* **2**: 299-322, 1760-1761.

FOURIER, Jean Baptiste Joseph. *Théorie analytique de la chaleur*. Paris: Firmin Didot, père et fils, 1822. Reprinted in: FOURIER, Jean Baptiste Joseph. *Oeuvres de Fourier*. Edited by Gaston Darboux. 2 vols. Paris: Gauthier-Villars et fils, 1888-1890. See specially 135-140.

FOURIER, Jean-Baptiste. Mémoire sur la statique contenant la démonstration du principe des vitesses virtuelles et la théorie des moments. Vol. 2, pp. 475-521, in: DARBOUX, Gaston (ed.). *Oeuvres de Fourier*. 2 vols. Paris: Gauthier-Villars et fils, 1888-1890.

GALILEI, Galileo. *Discorsi e dimostracioni matematiche intorno a due nuove scienze*. Vol. 13, in: ALBÈRI, Eugenio; BIANCHI, Celestino (eds.). *Le opere di Galileo Galilei*. 15 vols. Firenze: Società Editrice Fiorentina, 1842-1856.

GALILEI, Galileo. *Discorso intorno alle cose che stanno in su l'acqua o che in quella si muovono*. Vol. 12, in: ALBÈRI, Eugenio; BIANCHI, Celestino (eds.). *Le opere di Galileo Galilei*. 15 vols. Firenze: Società Editrice Fiorentina, 1842-1856.

GENOCCHI, Angelo. Dei primi principii della meccanica e della geometria in relazione al postulado d'Euclide. *Annuario* della *Società Italiana delle Scienze*, [3] **2**: 153-189, 1869 (a).

GENOCCHI, Angelo. Intorno ad una dimostrazione di Daviet de Foncenex. *Atti della Reale Accademia delle scienze di Torino* **4**: 323-327, 1869 (b).

GENOCCHI, Angelo. Sur un mémoire de Daviet de Foncenex et sur les géométries non-euclidiennes. *Memorie della*

Reale Accademia delle Scienze in Torino, [2] **29**: 365-389, 1878.

GENOCCHI, Angelo. Note biographiche intorno ai tre fondatori della R. Accademia delle Scienze. Pp. 86-95, in: CALAMIDA, Umberto (ed.). *Il primo secolo della R. Accademia delle Scienze di Torino.* Torino: Stamperia Reale di G. B. Paravia, 1883.

GOODWIN, Harvey. On the connexion between the sciences of mechanics and geometry. *Transactions of the Cambridge Philosophical Society*, **8**: 269-277, 1847.

GORRESIO, Gaspare. Notizia storica della R. Accademia delle Scienze di Torino. Pp. 3-6, in: CALAMIDA, Umberto (ed.). *Il primo secolo della R. Accademia delle Scienze di Torino.* Torino: Stamperia Reale di G. B. Paravia, 1883.

GRIGORIAN, Aszot Tigranowicz. Les travaux sur la mécanique non-euclidienne en Russie. *Scientia*, **95**: 347-350, 1960.

HANKINS, Thomas L. *Jean d'Alembert: science and the Enlightenment.* Oxford: Clarendon Press, 1970.

HANKINS, Thomas L. Response. *Isis*, **67**: 276-278, 1976.

HEATH, Thomas Little. *The thirteen books of Euclid's Elements.* 2nd. edition. 2 vols. New York: Dover, 1956.

HIGGINS, Thomas J. Electroanalogic methods. *Applied Mechanics Review*, **10**: 331-335, 443-448, 1957.

LAGRANGE, Joseph-Louis. *Mécanique analytique.* Vols. 11-12, in: SERRET, Joseph-Alfred (ed.). *Oeuvres de Lagrange.* 14 vols. Paris: Gauthier-Villars, 1867-1892.

LAGRANGE, Joseph-Louis. Recherches sur la nature et la propagation du son. *Mélanges de Philosophie et Matématique de la Société Royale de Turin* **2**: 39-148, 151-316, 319-332, 1760-1761.

LANGHAAR, Henry L. *Dimensional analysis and theory of models.* New York: John Wiley & Sons, 1951.

LAPLACE, Pierre Simon de. Recherches sur l'integration des équations différentielles aux différences finies et sur leur usage dans la théorie des hasards. Vol. 8, pp. 69-197, in P.

S. Laplace, *Oeuvres complètes de Laplace*. 14 vols. Paris: Gauthier-Villars, 1878-1912.

LARMOR, Joseph. Units, dimensions of. Vol. 27, pp. 736-738, in *Encyclopaedia Britannica*. 13th edition. 32 vols. Chicago: Encyclopaedia Britannica, 1926.

LEDIEU, Alfred. De l'homogeneité des formules. *Comptes Rendus Hebdomadaires des Séances de l'Académie des Sciences de Paris*, **96**: 1692-1696, 1883.

LEGENDRE, Adrien Marie. *Éléments de géométrie*. Paris: F. Didot, an II (1794).

MACAGNO, Enzo O. Historico-critical review of dimensional analysis. *Journal of the Franklin Institute*, **292**: 391-402, 1971.

MARTINS, Roberto de Andrade. The origin of dimensional analysis. *Journal of the Franklin Institute*, **311**: 331-337, 1981.

MARTINS, Roberto de Andrade. A busca da ciência a priori no final do século XVIII e a origem da análise dimensional. Pp. 391-402, in: MARTINS, Roberto de Andrade; MARTINS, Lilian Al-Chueyr Pereira; SILVA, Cibelle Celestino; FERREIRA, Juliana Mesquita Hidalgo (eds.). *Filosofia e História da Ciência no Cone Sul: 3° Encontro*. Campinas: AFHIC, 2004.

MONTUCLA, Jean Étienne. *Histoire des mathématiques*. 2nd. edition. 2 vols. Paris: [s.n.] 1802.

MORGAN, Augustus de. On the general principles of which the composition or aggregation of forces is a consequence. *Transactions of the Cambridge Philosophical Society*, **10**: 290-304, 1864.

NEWTON, Isaac. *Mathematical principles of natural philosophy*. Chicago: Encyclopedia Britannica, 1952.

PALACIOS, Julio. *Análisis dimensional*. Madrid: Espasa-Calpe, 1956.

PIONCHON, Joseph. Théorie des mesures: introduction a l'étude des systèmes de mesures usités en physique. Paris: Gauthier-Villars et fils, 1991.

POISSON, Siméon Denis. *Traité de mécanique.* 2 vols. Paris: Courcier, 1811.

POISSON, Siméon Denis. *Traité de mécanique.* 2nd. edition. 2 vols. Paris: Bachelier, 1833.

RAVETZ, Jerome. The representation of physical quantities in 18th century mathematical physics', *Isis,* **52**: 7-20, 1961.

RIABOUCHINSKY, Dimitri Pavlovitch. Méthode des variables de dimension zéro et son application en aérodynamique. *Aérophyle,* **1**: 407-408, Sept. 1911.

ROSENFELD, Boris Abramovich; ČERNOVA, M. L. Algebraic exponents and their geometric interpretation. *Organon,* **4**: 109-112, 1967.

SÉDOV, Leonid Ivanovich. *Similitude et dimensions en mécanique.* Moscou: MIR, 1977.

WHEWELL, William. On the nature of the truth of the laws of motion. *Transactions of the Cambridge Philosophical Society,* **5**: 149-172, 1834.

THE EARLY HISTORY OF DIMENSIONAL ANALYSIS: II. LEGENDRE AND THE POSTULATE OF PARALLELS

Roberto de Andrade Martins

Abstract: In an attempt to prove Euclid's fifth postulate, Legendre applied the principle of homogeneity. This was apparently the second historical use of dimensional analysis. This paper describes and analyses Legendre's work. It is shown that he assumed as a basis of his argument the inexistence of an absolute length standard – as Lambert did before him. Some reactions against and for Legendre's ideas are presented. It is shown that Legendre's work was premature: his method did not have a solid foundation, and the mathematicians of his time had good reasons to dismiss his arguments as invalid.
Keywords: dimensional analysis; postulate of parallels; foundations of geometry; non-Euclidian geometry; *a priori* proof; Legendre, Adrien-Marie

1. INTRODUCTION

A previous paper[1] has studied what seems to be the oldest instance of use of dimensional analysis in science: the

---

[1] See the previous chapter in this volume: Roberto de Andrade Martins. The early history of dimensional analysis: I. Foncenex and the composition of forces

MARTINS, Roberto de Andrade. *Studies in History and Philosophy of Science I*. Extrema: Quamcumque Editum, 2021.

demonstration of the law of force composition, in the Turin paper (Foncenex, 1760-1761) ascribed to François Daviet de Foncenex (1734-1798). The second episode in the history of dimensional analysis that came to our attention was the use of dimensional reasoning by Adrien-Marie Legendre (1752-1833) to derive the fifth postulate of Euclidian geometry, and other geometrical relations. Exactly as in the case of the Turin paper, here again the motivation was the attempt to provide an *a priori* proof of a fundamental law.[2]

From our contemporary point of view, we could perhaps say that there is a very deep difference between the two attempts, since the first one deals with a physical law, and the second one with a mathematical law. However, that difference was not altogether clear in the given historical context (France, in the last decades of the 18th century). Hence, we shall consider the two instances to be on the same footing, although nowadays no one even considers the possibility of applying dimensional analysis to mathematics itself.

We shall begin by studying the historical context of Legendre's work, and then display his ideas and the strong reaction that they have produced. Legendre's attempts to prove the postulate of parallels are part of the history that preceded the development of non-Euclidian geometry. However, since the rise of non-Euclidian geometry is well known and has deserved many historical studies, this part of the subject will be described here as briefly as possible.[3]

---

[2] The paper published here was written in 1981 but it has not been published until now. No attempt was made to improve or to update the content of the article. Only slight amendments were made.

[3] There are many books and papers on non-Euclidian geometry and on the history of mathematics that present an account of the subject. A very good report, although somewhat out of date, is the book by Roberto Bonola (1955). A collection of original texts may be found in Engel & Stäckel (1895) or in Sjöstedt (1968) – in this second book, both in the original language and translated into *interlingue*. Two very useful bibliographies describing most works published up to the

2. EUCLID'S FIFTH POSTULATE

The famous postulate of parallels of Euclidian geometry is sometimes described in modern textbooks in this way: "Through a given point can be drawn only one parallel to a given line". Actually, this is Proclus' or Playfair's axiom,[4] which is sometimes substituted for Euclid's original formulation: "That, if a straight line falling on two straight lines makes the interior angles on the same side less than two right angles, the two straight lines, if produced indefinitely, meet on that side on which are the angles less than the two right angles" (Heath, 1952, p. 2). Actually, since Euclid's postulate has always seemed strange and cumbersome to most geometers, several authors have tried to eliminate it or to substitute this postulate by another simpler and more 'intuitive' assumption, Among the many attempts we shall describe those that are directly relevant for the understanding of Legendre's work.

John Wallis (1616-1703) has suggested the following alternative axiom: "To every figure there exists a similar figure of arbitrary magnitude" (Wallis, 1693, vol. 2, pp. 674-677; Sjöstedt, 1968, pp. 83-95; Engel & Stäckel, 1895, pp. 21-30). From one particular instance of this axiom − the existence of *similar triangles*, with equal angles but different sizes − he derived a correct proof of Euclid's postulate of parallels. Since the existence of similar figures of different sizes seemed to him more intuitive than the postulate of parallels, he proposed to substitute the latter by the former assumption.[5]

---

beginning of the 20th century have been used: Sommerville (1911) and Loria (1931).

[4] This form of the postulate is equivalente to Euclid's proposition 31 of Book I, which was a theorem, but which was later used as an axiom (Wolfe, 1945, pp. 20-21).

[5] To several mathematicians, since the early 19th century, this seemed the most natural axiom which should be used, instead of Euclid's original formulation. See Carnot (1803, p. 481), Laplace (1878-1912, vol. 6, pp. 471-472; Delboeuf, 1895; Hill, 1927).

A second approach that interests us is the one developed by Giovanni Girolamo Saccheri (1667-1733) and by Johann Heinrich Lambert (1728-1777). Studying a quadrilateral (Fig. 1) where *AD* and *BC* have equal lengths, and where *A* and *B* are right angles, Saccheri discussed the question: can we prove, without assuming the postulate of parallels, that angles *C* and *D* are right angles? (Saccheri, 1733; Halsted, 1920; Sjöstedt, 1968, pp. 96-176; Smith, 1929, pp. 351-382; see also Beltrami, 1889). He showed that this cannot be proved, and that there are three alternatives: (i) *C* and *D* are acute angles; (ii) *C* and *D* are right angles; or iii) *C* and *D* are obtuse angles. There are no other alternatives, because Saccheri proved that both angles must be equal. Depending on the choice between these alternatives, one may derive that the sum of the angles of a triangle is respectively less then, equal to, or greater than two right angles. The hypothesis of the right angle leads to Euclidean geometry, because if the other postulates and axioms of Euclidian geometry are assumed, than the sum of' the angles of a triangle are equal to two right angles if and only if the postulate of parallels holds.

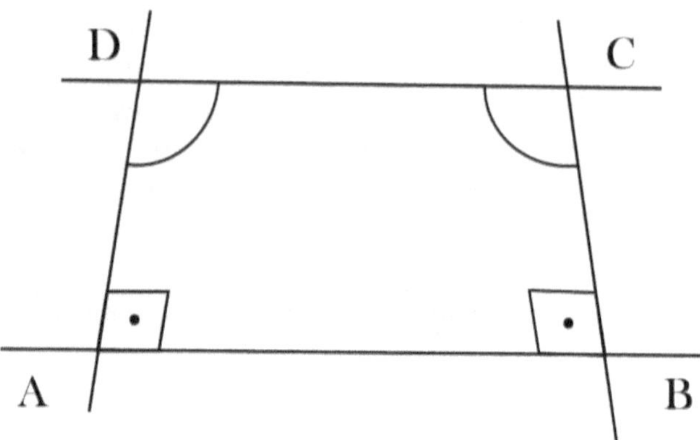

Fig. 1 – Saccheri's quadrilateral

Lambert has developed a series of ideas similar to those of Saccheri, and furthermore he obtained some important new conclusions (Lambert, 1786; Engel & Stäckel, 1895, pp. 135-208; Sjöstedt, 1968, pp. 177-250). If the hypothesis of the acute angles is true, then all triangles will have the sum of their angles less than two right angles, *but this sum will be different for different triangles*. The difference between this sum and two right angles will be *proportional to the area of the triangle*.[6] This implies that it is impossible to build two similar triangles of different sizes, and, generally, it will be impossible to build two similar figures of different sizes. Equilateral triangles with different sizes will have different angles, and there will be a one-to-one correspondence between lengths and angles less than 60°.

It follows from Lambert's work that lengths have an *absolute* significance, if the hypothesis of the acute angle is true, since figures of different sizes will have different properties. We might build absolute length standards, on the hypothesis of the acute angle, by choosing as unit, for instance, the side of an equilateral triangle such that the sum of its angles is equal to one right angle (Bonola, 1955, pp. 46-49).

Since, according to the conception of space that was current in the 18th century, it was impossible to have an absolute standard of length, one should deny the hypothesis of the acute angle, as Lambert did, and derive the postulate of parallels.

Now we can discuss Legendre's ideas.

3. LEGENDRE'S PROOFS OF THE POSTULATE OF PARALLELS

In the several editions of his *Éléments de géométrie*, from 1794- onward, Legendre attempted to prove the postulate of parallels in different ways. It seems that he was deeply concerned with this problem all through his life. In an article he published in the year of his death (1833), Legendre discussed

---

[6] The same consequence holds if the obtuse angle hypothesis is accepted. In both cases, for very small triangles, the difference between the sum of the angles and two right angles becomes negligible.

this subject for the last time, reviewing all his former attempts and presenting a new formulation of some proofs.

It is sometimes said that he added nothing new to the attempts of his predecessors, and that only his style of presenting the proofs was new. In the specific instance to be discussed here, it will be seen that this is a too negative evaluation, and it would not be shared by the mathematicians of Legendre's time.

Most of Legendre's demonstrations have a purely geometric style, but one of them – the one that interests us here – is grounded upon an *analytic* argument. It was published in the first edition of his *Elements*, in 1794, and abandoned for other methods in some editions of his book;[7] but it reappeared in latter editions, and it was presented again in Legendre's last article (Legendre, 1833; Sjöstedt, 1968, pp. 251-325). This proof has been usually neglected by historians of mathematics. Let us reproduce part of this demonstration.

> It is immediately demonstrated by superposition, and without any preliminary proposition, that *two triangles are equal when they have an equal side adjacent to two angles respectively equal.* Let us call this side p, the two adjacent angles A and B, the third angle C [Fig. 2]. It is therefore required that the angle C be entirely determined, when the angles A and B, with the side p, are known; for, if several angles C could correspond to the three given quantities A, B, p, there would be a corresponding number of different triangles that would have an equal side adjacent to two equal angles, which is impossible; hence the angle C must be a determined function of the three quantities A, B, p; and this I express thus, $C = \varphi:(A, B, p)$.

---

[7] The first edition is Legendre, 1794. The following editions are: 1796 (2nd), 1800 (3rd), 1802 (4th), 1804 (5th), 1806 (6th), 1808 (7th), 1809 (8th), 1812 (10th), 1817 (11th), 1823 (12th). There were some subsequent versions which reproduce the 12th edition, sometimes without the notes at the end of the book, where Legendre's analysis of the postulate or parallels can be found.

Let the right angle be equal to unity; then the angles A, B, C, will be numbers comprised between 0 and 2; and since $C = \varphi:(A, B, p)$, I say that the line p cannot enter into the function φ. Indeed, we have seen that C must be entirely determined by the given A, B, p alone, without any other angle or line; but the line p is heterogeneous with the numbers A, B, C; and if there is any equation among A, B, C, p, one might obtain the value of p from A, B, C; hence it would follow that p is equal to a number, which is absurd; hence p cannot enter into the function φ, and we have simply $C = \varphi:(A, B)$.

This formula already proves that if two angles of one triangle are equal to two angles of another [triangle], the third angle of the former must also be equal to the third angle of the other; and this being conceded, it is easy to work out the theorem we had in view. (Legendre, 1794, pp. 287-288; 11th edition, p. 281; 12th edition, p. 281; Legendre, 1833, pp. 372-373).

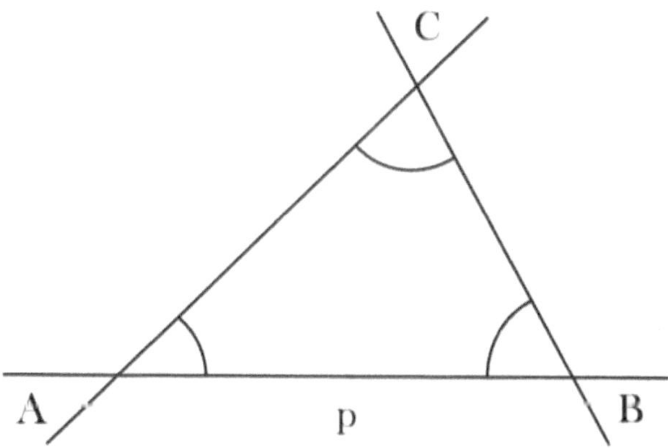

Fig. 2 – The triangle studied by Legendre.

At this point, Legendre was trying to prove the existence of similar triangles, independently of their sizes. If this is proved, then it is straightforward to prove both the right angle hypothesis and the postulate of parallels. In order to prove the existence of similar triangles, he tries to prove that the third angle of any triangle cannot depend on the side of the triangle, it can only be a function of the two other angles. This had been assumed by Wallis, but Legendre attempted to provide a proof of this assumption; and here he used an analytic argument which is just the assumption of the dimensional homogeneity of formulae.

We may derive Legendre's argument from the following assumptions, some of which were not explicitly stated by him:

AML 1 – Angles are geometrical magnitudes of zero dimension – that is, they are pure numbers.

AML 2 – A mathematical function of pure numbers can only yield pure numbers.

AML 3 – A line is not a pure number, and it cannot be represented by pure numbers in the same way as angles.

AML 4 – Heterogeneous quantities cannot be equal.

AML 5 – In the equation $C = \varphi:(A, B, p)$ there is only one quantity with a geometrical dimension, and this is the line p.

If those analytical conditions are accepted, then Legendre's proof is correct. But not all of those assumptions have been accepted by everyone, as will be seen below.

In a footnote to the text transcribed above, added in later editions, Legendre answered a first objection:

> It has been objected against this demonstration that if it were applied, word by word, to spherical triangles, it would result that two known angles are sufficient to determine the third, which does not happen in this kind of triangles. The answer is that in spherical triangles there is one element more than in plane triangles, and this element is the radius of the sphere, which must not be forgotten. Let r be the radius, then, instead of $C = \varphi:(A, B, p)$, we shall have $C = \varphi:(A, B, p, r)$, or just $C = \varphi:(A, B, p/r)$, by the law of homogeneity. But, since the ratio p/r is a number, such as A, B, C, nothing hinders p/r

from being found in the function, and hence one cannot conclude that $C = \varphi{:}(A, B)$. (Legendre, 1800, p. 312)

Let us ponder upon this objection. In any given spherical surface, a spherical triangle is determined by one side p and two adjacent angles A and B. If we are always referring to this same spherical surface, we may say that the third angle C is a function of the three variable parameters A, B, p. But when we try to put this function $C = \varphi{:}(A, B, p)$ into an algebraic form, we notice that we must introduce a constant parameter r which has the same dimensions as the side p. This does not introduce a new variable, but allows us to transform one of the former variables (p) into a pure number, dividing the side p by the dimensional constant r.

We shall now reproduce the same argument introducing a small change.

Let us think about our universe. In our given universe, a triangle is determined by one side p and two adjacent angles A and B. Since we are always referring to this same universe, we may say that the third angle C is a function of the three variable parameters A, B, p. But when we try to put this function $C = \varphi{:}(A, B, p)$ into an algebraic form, we notice that we must introduce a constant parameter r which has the same dimensions as the side p. This does not introduce a new variable, but allows us to transform one of the former variables (p) into a pure number, dividing the side p by the dimensional constant r.

Legendre assumed that this parameter r cannot exist in our actual space, and this is expressed in the fifth assumption (AML 5). The existence of a constant parameter r would mean that our universe has a determined length scale, or that there is a natural length standard; lengths would have an absolute meaning – which is exactly what Lambert denied in order to prove the postulate of parallels. Hence, the dimensional argument employed by Legendre is, at its bottom, equivalent to Lambert's demonstration – but this equivalence was not noticed at that time.

We therefore see that Legendre's implicit assumption of the inexistence of geometrical dimensional constants is a fundamental step that allowed him to derive the postulate of parallels. In non-Euclidian geometries, this assumption is not true, of course.[8]

This is a very general and important warning regarding any use of dimensional arguments: we must always seek for the existence of dimensional constants related either to the specific system that is being studied, or to our universe. If we do not take into account all the dimensional quantities on which the required result may depend, the method of dimensional analysis will produce wrong results.[9]

Using the same method Legendre also derived other Euclidian theorems from homogeneity conditions. He proved that "in equiangular triangles, the sides opposite to equal angles are proportional"; that "in similar figures the homologous lines are proportional"; that "the surfaces of similar figures are to each other as the squares of the homologous sides"; and several other related theorems about similar solids, circles, spheres – theorems that are valid only in Euclidian geometry. In those demonstrations he assumed that surfaces and volumes are respectively homogeneous with the squares and cubes of lengths, as was usually accepted.

The source of Legendre's method is explicitly stated at the end of the Note that contains these arguments – it is the Turin paper:

---

[8] In the context of non-Euclidian geometry, such a parameter does indeed appear; and it seems that the simplest description of our physical universe is one that assumes a non-Euclian space with a curvature radius of about 10^{10} light-years (Einstein, 1951, appendix I; Robertson, 1933; Lemaître, 1949; Rindler, 1967; Sandage, 1970; Barrow, 1978).

[9] At Legendre's time, this was not known. See, however, Lord Rayleigh (1945, vol. 1, pp. 54-55).

> We finally remark that the consideration of functions, which thus affords a very simple demonstration of the fundamental propositions of Geometry, has already been employed with success in the demonstration of the fundamental principles of Mechanics. See the *Mémoires de Turin*, volume II. (Legendre, 1794, p. 294; 11th edition, p. 286; 12th edition, p. 287).

This remark shows that Legendre knew the work ascribed to Foncenex, and he accepted it as valid.[10] Since Legendre did not refer to any other previous use of the same method by mathematicians, it seems that nobody tried to apply the principle of homogeneity to geometrical theorems, before Legendre – although this use had already been suggested in the Turin paper, as has been shown.[11]

The novelty of the use of dimensional arguments in geometry is also implied by a commentary on Legendre's work by Baron Jean Frédéric Théodore Maurice (1775-1851):

> The algorithm of functions had been previously employed with success in establishing the fundamental principles of Mechanics (see the *Miscell. Taurin.* vols. I and II); but this new application does not yield to those which have gone before. (Maurice, 1819)

It might seem that the plural form 'those' ("celles qui l'ont précédée") implies that Maurice knew several applications of the principle of homogeneity earlier than Legendre's work. But this may also be interpreted as referring to the several uses in

---

[10] Joseph Pionchon reproduced in his book Legendre's dimensional argument, copied from the 12th edition of the *Éléments* (Pionchon, 1891, pp. 228-234). However, he did not reproduce the two last paragraphs of the Note of that edition, where Legendre mentioned the Turin paper. Pionchon's presentation was written so as to imply that Legendre was the first author to use the principle of homogeneity in *a priori* proofs of formulae.

[11] See my previous paper, in this volume.

the Turin paper itself; or to the uses of dimensional arguments to check equations. Anyway, Maurice seems to be aware of no previous use of dimensional reasoning to prove geometrical principles.

4. REACTIONS TO LEGENDRE'S ARGUMENT

It seems that Legendre's dimensional proof of the postulate of parallels has produced little or no reaction in France. At the beginning of the 19th century, French mathematicians believed that Euclid's Geometry was true and unique, and that the fifth postulate could be proved; hence, Legendre's proof was accepted as correct without much discussion as one among several other equivalent proofs.[12]

The first important criticism of the method came from John Leslie (1766-1832), professor of mathematics at the University of Edinburgh, who had produced his own proof of the fifth postulate. In a Note to the second edition of his *Elements of geometry*, Leslie described Legendre's argument, and he commented:

> To a speculative mathematician this argument is very alluring, though it will not bear a rigid examination. Many quantities in fact appear to result from the combined relations of other quantities that are altogether heterogeneous. Thus, the space which a moving body describes, depends on the joint elements of time and velocity, things entirely distinct in their nature; and thus, the length of an arc of a circle is compounded of the radius, and of the angle it subtends at the center, which are obviously heterogeneous magnitudes. For aught we previously knew to the contrary, the base c [or p, on

---

[12] In his last paper on this subject, Legendre stated that no objections had been raised against the correctness of his proofs, but only to their complexity; and he tried in that paper to select his simplest demonstration of the postulate of parallels (Legendre, 1833, pp. 407-408). This was not literally correct, but it seems close to the truth, as regards the French mathematicians.

the triangle studied by Legendre] might, by its combination with the angles A and B, modify their relation, and thence affect the value of the vertical angle C. In another parallel case, the force of this remark is easily perceived. Thus, when the sides a, b and their contained angle C are given, the triangle is determined, as the simplest observation shows. Wherefore the base c is derived solely from these data, or $c = \varphi{:}(a, b, C)$. But the angle C, being heterogeneous to the sides a and b, cannot coalesce with them into an equation, and consequently the base c is simply a function of a and b, or it is the necessary result merely of the other two sides. Such is the extreme absurdity to which this sort of reasoning would lead! (Leslie, 1811, pp. 403-404)

This quotation shows that Leslie has misunderstood Legendre's method. If Legendre's assumption amounted just to say that we cannot combine heterogeneous quantities into a single function, then Leslie's criticism would be just. If Legendre did not assimilate angles to pure (abstract) numbers, Leslie's argument would also be correct. But Leslie did not perceive this important difference that Legendre established between angles and lines, and this was fundamental to his argument. Hence, Leslie's criticism was unfair.

It seems that the main reason for Leslie's attack against Legendre has been that the latter tried to provide an *a priori* demonstration of the fifth postulate:

The profound geometer already quoted, pursuing his refined argument, has, from the consideration of homogeneous quantities, likewise attempted to deduce the proportionality of the sides of equiangular triangles. But in this abstruse research, assumptions are still disguised and mixed up with the process of induction. Such indeed must be the case with every kind of reasoning on mathematical or physical objects, which proceeds *a priori*, without appealing, at least in the first instance, to external observations. Of this kind, are some of those ingenious analytical investigations respecting the laws of motion and the composition of forces. (Leslie, 1811, pp. 405-406)

I have quoted this passage to show the specific aversion that Leslie (and some other scientists and mathematicians from Great Britain) showed as regards *a priori* proofs – an attitude very different from that of French mathematicians of that time.[13] However, there was another Edinburgh mathematician who did not have the same opinion: John Playfair (1748-1819). In a review of a French book (Delambre, 1808) where Legendre's analytical proof of the 5th postulate was described, Playfair disclosed a positive evaluation of this method (Playfair, 1810).[14] He first criticized Legendre's *geometrical* proof as too complex, and then he remarked:

> The other demonstration, however, which is in the Notes, possesses the most perfect simplicity, at the same time that it is new; proceeding on a principle that has been long recognized, but from which no consequence, till now, has ever been deduced. (Playfair, 1810, p. 3)

Shortly after the publication, in 1811, of the second edition of Leslie's *Elements of geometry*, Playfair reviewed this book, and criticized Leslie's remarks regarding Legendre's method (Playfair, 1812).[15] After praising again Legendre's argument, Playfair showed that Leslie had misunderstood this proof

---

[13] This was a common though not an universal attitude of British mathematicians, and was remarked by one of them, in an anonymous letter to the Editor of the *Philosophical Magazine* (Sigma, 1825). In this article, Sigma referred especially to papers in the *Edinburgh Review* and of the *Encyclopaedia Britannica*, where the French author was attacked. Although Sigma himself seems to support Legendre, his interpretation concerning the grounds of his proofs is completely different from the argument presented by the French mathematician.

[14] The paper was published anonymously, but its author was identified by Leslie.

[15] The review was also published anonymously.

(Playfair, 1812, pp. 89-91), presenting the reasons which have been shown above.

I quote below Playfair's elucidation of Legendre's argument, since his explicitly used the word 'dimension' – avoided by Legendre – and throws a richer light on the argument:

> The quantities A, B, C are angles; they are of the same nature as numbers, or mere expressions of ratio, and, according to the language of algebra, are of no dimension. The quantity c, on the other hand, is the base of a triangle, that is to say, a straight line, or a quantity of one dimension. Of the four quantities, therefore, A, B, C, c, the first three are no dimensions, and the fourth or last is of one dimension. No equation therefore can exist involving all these four quantities, and them only; for if it did, a value of c might be found in terms of A, B, and C, and c would therefore be equal to a quantity of no dimension; which is impossible. It would be equal to a quantity of no dimensions, because every function of quantities of no dimension, must itself be of no dimension. (Playfair, 1812, p. 90)

In the above quotation, Playfair explicitly presented some of Legendre's implicit assumptions. The presentation is different, but the two arguments are equivalent.

5. LESLIE'S SECOND ATTACK

Besides Playfair's article, Leslie's criticism was answered by Legendre himself, who wrote a private letter to him. And although Leslie's first charge against Legendre had certainly been refuted, he launched a new attack, in the third edition of his book, in which he reproduced an extract from Legendre's letter (Leslie, 1817, p. 296), and reiterated his earlier criticism, before adding:

> The whole stress of the argument, it may be perceived, lies in the distinction which M. Legendre endeavors to establish between angles and lines – a distinction which I hold at

bottom to be merely arbitrary. Angles and lines are both equally real quantities, though of different kinds; they are capable of being measured, and consequently represented by numbers, by referring each of them to some definite measure or unit of its own denomination. Angles are measured or expressed numerically by angles, and lines by lines. It is true that the mensuration of angles is facilitated by a reference to a subdivision of the circuit or entire revolution; yet even this mode of denoting angular magnitude is evidently only conventional. As standards for measuring straight lines, nature has furnished the limbs of the human body, and the extent of our globe itself. Such units of mensuration are not indeed very definite or readily attainable; but they are not therefore the less real or prominent. Nor is there any essential difference in principle between the expressing of an angle by degrees, of which 360 or 400 are contained in a complete revolution, and the denoting of a straight line on the French system, for instance by the number of meters it includes, each of which is the forty millionth part of the entire circumference of the earth. Angles and lines hence present to the mind no radical or absolute discrimination, and therefore the argument grounded on such a distinction must lose all its efficacy. (Leslie, 1817, p. 297)

Leslie added that, except for a single exception, he was acquainted with no geometer of any eminence in Britain, who did not admit the fallacy of the argument employed by Legendre. The exception was, obviously, Playfair.

In support of his position, Leslie also quotes part of a letter he received from a mathematician whose name he omits, but who was later identified as James Ivory (1765-1842). Leslie referred to him as the head of the British mathematicians and reproduced his criticism, that amounted to show that Legendre's argument presupposes a geometrical hypothesis equivalent to Euclid's fifth postulate (Leslie, 1817, pp. 294-295).

Leslie's new objection is very strong, indeed. Angles are not abstract numbers. They may be reduced to numbers if we divide them by some unit – but the same holds for any concrete magnitude. What exactly allows us to distinguish the unit of angles as *natural*, and the unit of length as *arbitrary*? That was not a very clear issue, at that time.

Two French answers to Leslie's new argument have been produced: one by baron Jean Frédéric Théodore Maurice (1775-1851) and the other one by Legendre himself.

In his paper, Maurice first tried to elucidate the principle of homogeneity (Maurice, 1819). His argument is not altogether clear, but I think that his ideas are as follows. Arithmetic and algebra only state relations between abstract numbers; Any formula of geometry or physics must be constructed in such a way that it may be reduced to a relation between abstract numbers, by dividing each term of the equation by the same unit. The equations of mechanics seem to relate concrete magnitudes, such as spaces and times, but they are really equations between abstract numbers, since it is always supposed that a line-unit and a time-unit are necessarily included in the equations, so that, every length being divided by its unit, and each time by its unit, only abstract numbers remain.

This view was indeed assumed by most physicists of that time, as we have already shown;[16] this was exactly the principle of homogeneity used by Poisson, and it is not equivalent to or compatible with the ideas used by Legendre,

After some general remarks, Maurice returned to the geometrical problem, stating:

> Consequently, in the case before us, when we have arrived at the general relation expressed by the symbolical equation $C = \varphi{:}(A, B, c)$, it is *rigorously essential* to its existence that it be capable of being reduced to a relation among abstract numbers. Now, if only the angles A, B, C entered into it, there would be no difficulty: for since each of them expresses a multiple or submultiple of the angular unit, this unit may be made to disappear by means of division. But if the straight-

---

[16] See my previous paper in this volume, on Foncenex and the composition of forces.

line *c* enters also, this relation becomes manifestly absurd, since it contains two heterogeneous units, which cannot both be made to disappear from the calculation. [...] Thus, it is because the line *c* is *the only* straight line which occurs in the proposed relation, that we are rigorously authorized *a priori* to eliminate this line from it, as a quantity which cannot remain without leading to a manifest absurdity. (Maurice, 1819, p. 90)

Maurice's argument is not equivalent to the one used by Legendre, but it is exactly the one properly criticized by Leslie in his first attack. In Legendre's argument, the essential point is that *angles are abstract numbers*, and Maurice did not use this assumption here. Hence, the elucidation presented by Maurice does not provide an answer to Leslie's criticism.

However, Maurice added some different remarks. He tried to show that a relation between the sides of a triangle and one of its angles is not absurd, although here we have only one angular magnitude. What is the difference between this case and that of the angles of the triangle and one of its sides? Maurice did not state that angles are numbers; but that a transcendental function of the angle (such as *cos C*) might appear in a relation such as $c = \varphi:(a, b, C)$; and this transcendental function is a number.

But why can we find a function that produces an abstract number from a single angular magnitude, and we cannot find a function that produces an abstract number from a single *linear* magnitude? What is, after all, the difference between angles and lines? Maurice tried to ascertain the difference:

These quantities (angles and lines) regarded as *magnitudes* destined to enter into our calculations, are not homogeneous, when referred to the wholes of which they respectively are parts. The angle is a portion of a finite whole, the straight line is a portion of an infinite whole; so that *every given angle is a finite quantity*, whilst *every given straight line is a quantity infinitely small*, and *only the ratios of given straight lines can enter into our calculations with given angles*. (Maurice, 1819, p. 92)

This was an *ad hoc* argument that did not have any sound justification. It seems that Maurice used here a hypothesis equivalent to: "Two magnitudes are homogeneous if and only if they are both finite or infinitely small relative to the wholes from which they are parts". If we accepted this assumption, we could show that lines and times are homogeneous quantities, since they are infinitely small when referred to the infinite wholes of which they form part. If this criterion of homogeneity was accepted, all magnitudes would be split in just two classes: time, length, surface and volume would be homogeneous; and they would be heterogeneous to the class that includes plane angles, solid angles, etc. But this difference is not altogether clear. We can think about angles corresponding to several revolutions – such as the angle described by the Earth during one year – and there is no upper limit to the value of an angle, if this is conceded. Besides that, we measure time by the angle described by the hands of our analogic clocks; both angles and times have some cyclic properties, but both of them may be regarded as unlimited or infinite.

Let us consider another border-line case. If we take into account curved lines, such as an arc of a circumference, this line may be regarded as part of a finite whole. Are curved lines homogeneous to angles? Are curved surfaces homogeneous to curved lines? It seems that Maurice did not take into account all these consequences of his assumption. He was trying to justify Legendre's method and to criticize Leslie, and any *ad hoc* argument seemed useful, even if it did not agree with Legendre's own ideas and if it had no support.

At another point of his article, Maurice criticized the analogy that Leslie proposed between the unit of angular quantities and the unit of length. There is one *natural* angular standard: the whole revolution, which amounts to four straight angles. But there is no natural unit of length, since the straight line is infinite, and hence any unit of length is arbitrary (Maurice, 1819, p. 94).

In order to support Legendre's argument, it is indeed necessary to show that we can represent angles by abstract numbers in a natural way but that the same cannot be done with lengths. Maurice could have done this, but he did not complete the argument; it seems, indeed, that he did not accept that angles are abstract numbers. Hence, his answer to Leslie missed the fundamental point.

Besides, Maurice's distinction between 'natural' angular units and 'artificial' or 'arbitrary' units of length was not altogether clear. First: we may have several different 'natural' angular units, such as a complete revolution, or the straight angle, or the angle of an equilateral triangle in Euclidian geometry, or the radian. If we do not specify which of these natural units we are using, we cannot know what does it mean that some angular quantity amounts to 0.3 or 2.7, or any other value. So, angles are not indeed pure numbers. Besides, we may have 'natural' length standards if the postulate of parallels is not true, as had been shown by Lambert; and since the whole argument was built by Legendre exactly to prove that postulate, one cannot just assume that there are no natural units of length.

All this shows that Maurice did not understand that the basic point in Legendre's argument was the equivalence between angles and abstract numbers; that he probably did not agree with this assumption; and that nevertheless he tried to defend Legendre's argument by some obscure considerations that do not touch the relevant difficulties. What is still more strange: Legendre gave his support to Maurice's defense, as will be shown below.

Up to the eleventh edition of his book, Legendre did not refer to Leslie's attack. However, in the 12th edition, he added to his Note II the following remark:

> Finally, although the aforementioned theory is established upon the most solid foundations, we should not conceal that it has been attacked by M. Leslie, a famous professor at Edinburgh, in his *Elements of Geometry*, second and third editions; but without entering into any detail of this subject, it

is enough to say that M. Leslie's objection have been completely refuted, first by M. Playfair, his countryman, in the *Edinburgh Review*, volume XX, and then by M. Maurice, of the Paris Academy of Sciences, in the *Bibliothèque Universelle de Genève*, October 1819. One can also see the discussion of these same objections in the English edition of our *Elements* given by M. David Brewster, Edinburgh, 1822. (Legendre, 1823, p. 287)

In the British edition, Legendre added to his Note II a translation of most of Maurice's article, thus seeming to endorse all his ideas (Legendre, 1822, pp. 230-238). At the point where Maurice stated that angles are finite quantities, and lengths are infinitely small quantities, Legendre added a footnote, remarking that "this is a very subtle and very just metaphysical idea: it is, at the same time, strictly analytical [...]" (*ibid.*, p. 235). I fail to see how can a 'strictly analytical' idea be, at the same time, 'metaphysical'.

Legendre's acceptance of Maurice's ideas is quite peculiar. Is it possible to believe that Legendre did not perceive the relevant elements of his argument? We may interpret Legendre's attitude as a strategic move. Legendre was trying to defend himself from Leslie's assault; Maurice appeared and offered his help against Leslie; Legendre accepted Maurice's aid, although the latter's ideas were neither correct nor equivalent to Legendre's original argument. The acceptance of Maurice's support seems unreserved, and Legendre praised even his nonsenses.

In the British edition of his book, Legendre also added his own defense against Leslie's attack (Legendre, 1822, pp. 227-230). The important point made by Legendre in this reply is again the distinction between angles – which he once more regarded as abstract numbers – and lengths; and the difference between the 'natural' angular unit (the right angle) and the arbitrary units of length. But enough has been said about this subject, and we shall not repeat the details of Legendre's vindication. Suffice it to say that this debate had the effect of drawing Legendre to the explicit use of the hypothesis of the inexistence of a natural unit of length.

In his last paper on the theory of parallels, Legendre again advanced the analytical argument (Legendre, 1833). After stating that angles are numbers, or may be represented by numbers, he added: if the third angle of a triangle depended on the base of the triangle, there would exist some way of computing this side from the knowledge of the three angles; and he remarked:

> But the absurdity of a result such as this is manifest; because the relation, whatever it might be, which will determine the side *AD* with the aid of the three numbers [...] cannot give for *AD* but a number. [...] If this number is 12, for instance, nothing can be derived from this about the absolute value of *AD*, because it would be required to know which unit of length is linked to the number 12 – whether they are millimeters, meters, feet, furlongs, leagues, etc. The nature of the question gives no light about this, it does not show which is the unit of length; and it is precisely the absence of any length unit that makes the above result absurd. We see that by our hypothesis one could retain forever a length measure taken as unity. It would suffice for that to keep the memory of three numbers (the angles of the triangles, or only two if the triangle *ADE* was supposed isosceles, or even one single, if it was supposed equilateral. (Legendre, 1833, p. 391)

Here, Legendre explicitly acknowledged that his proof was grounded on the supposition of the inexistence of a natural unit of length – the basic hypothesis used by Lambert in his proof of Euclid's fifth postulate, as was shown in Section 2 of this paper. Actually, the question of units is not so fundamental for the argument. If there is at least one special constant length that can enter into the formula, then there would be no absurdity in computing the side of the triangle from its angles.

6. REACTIONS PUBLISHED IN THE PHILOSOPHICAL MAGAZINE

A few papers that have appeared in the *Philosophical Magazine*, between 1822 and 1825, commented about Legendre's method. They were possibly a reaction against the publication in 1822 of the English translation of Legendre's book. They will be briefly described in this section.

The first two papers accepted Legendre's analytical argument, although they criticized and tried to improve some of its geometrical assumptions (Ivory, 1822; Meikle, 1822). They will not be discussed here. Shortly afterwards, John B. Walsh (1786-1847) criticized the analytic method (Walsh, 1824). Walsh described the weak point, pointing out that angles are not numbers and that the right angle is an arbitrary unit of angles, not a natural or unique unit. Exactly as angles may be transformed into numbers (by dividing them by the right angle or any other angular unit), exactly in the same way the side of the triangle may be transformed into a number, dividing it by a length unit – the meter or any other standard. Besides these appropriate remarks, Walsh added a lot or rhetoric, and some wrong arguments against Legendre.

The following article was published by an anonymous author, who signed the paper as 'Dis-iota'.[17] This author criticized the geometrical part of Legendre's argument, but accepted and defended the analytical method (Dis-iota, 1824a). In defense of the different roles of angles and lines in geometric equations, he wrote:

> In the problems of plane geometry, where lines and angles
> are combined in the same equations, the quantities depending

---

[17] 'Dis-iota" was James Ivory, the British mathematician cited by Leslie in support of his views (Leslie, 1817, 294-295). At the end of his paper, Dis-iota complained about Leslie's use of his letter, and remarked that Leslie had omitted several parts of the correspondence that were favourable to Legendre.

> on the angles invariably contain in their expressions nothing
> else but ratios, or the quotients of homogeneous magnitudes;
> which renders the equations independent of the manner in
> which the angles are themselves compared or measured. It is
> not the same with regard to lines; for the algebraic symbols of
> these always involve an arbitrary unit. (Dis-iota, 1824a, p.
> 162)

If Dis-iota meant that in geometrical equations angles only
appear together with and divided by other angles, as the above
quotation seems to imply, then Dis-iota was wrong. But perhaps
he meant that the angles may be represented by the ratio of the
arcs corresponding to the angles in a circle, divided by the
quadrant of the same circle, since he used this representation at
another point of his paper. But why should we choose the
quadrant (or the fourth part of the circumference) and not any
other different fraction of the circumference? The arc used for
comparison is arbitrary, and therefore Dis-iota's remark does
not solve the difficulty. In further remarks on the same subject,
he was unable to present a better argument (Dis-iota, 1824b).

Dis-iota incidentally criticized Walsh, who published a paper
to back his views (Walsh, 1824). Again, he presented his strong
criticisms to those who confound concrete quantities and
abstract number.

The next two articles were published by a new anonymous
correspondent, who signed his contributions as 'Sigma' (1825a,
1825b). After disclosing the identity of Dis-iota, whom he
identified as James Ivory (Sigma, 1825a, p. 101), he presented
his own criticism of Legendre's method. Although there are
some elementary mistakes in his article, it contains the most
lucid exposition of the correct form of the dimensional
argument, very similar to the one used nowadays. After denying
the validity of Legendre's comparison of angles to abstract
numbers, Sigma stated:

> We are [...] inclined to search for the cause of the
> difficulties that have startled us, in some imperfection in the

> mode of expressing or of treating the equations. That imperfection seems to me to be a deficiency in the original equation as given by Legendre. Superposition does not inform him thot the vertical angle (C) is determined by the base (c) and its adjacent angles (A and B) *alone* – but merely that it will be constant it they are constant; and hence that it must be determined by these *variables* and CONSTANTS alone. Noting these constants by γ, the original equation becomes
> $C = \varphi{:}(c, A, B, \gamma)$. (Sigma, 1825, p. 104)

However, as Sigma added, those constants must be either constant lines or constant angles; but since there are no constant lines, γ can only be a constant angle – the right angle, or any multiple or fraction of the right angle. Now, C depends on four quantities, but only one of them is a length. Applying the principle of homogeneity, we can now exclude this length c from the equation, which becomes:

$$C = \varphi{:}(c, A, B, \gamma).$$

Hence, Sigma improved Legendre's argument, denying the assumption that angles are abstract numbers (AML 1) and adding the correct assumption (in Euclidian geometry): there are some special or 'natural' angular magnitudes which may enter into the function, but there are no special or 'natural' lengths that may enter into the formula.

Unfortunately, I have been unable to find out the identity of Sigma – someone relevant in the history of dimensional analysis, since he was seemingly the first author who pointed out the importance of considering *constant magnitudes* in dimensional arguments.

7. NON-EUCLIDIAN GEOMETRY

It was to be expected that some criticism of Legendre's method would arise with the development of non-Euclidian geometry. Indeed, Nikolai Ivanovich Lobatschewsky (1792-1856) referred several times to Legendre, always disapprovingly (Bonola, 1955, p. 88; Gonseth, 1955, pp. vi-88, vi-102-114). In the introduction of his *Geometrical researches on the theory of parallels* (Lobatschewsky, 1866), for instance, Legendre's name appears three times concerning the attempts to prove the postulate of parallels, and no other geometer is cited:

> Legendre's efforts have added nothing to this theory [of parallels], since this author has been forced to leave the way of rigorous thought, throwing himself in circular considerations, and using principles that he attempts to exhibit as necessary axioms, without sufficient reasons. [...]
>
> The extension of this work [Lobatschewski's own researches] has perhaps hindered my countrymen from following this study, that, after Legendre, seemed to have lost its interest. But nevertheless, I still believe that the theory of parallels still deserves the attention of geometers, and it is for that reason that I propose myself to expose here what is essential in my researches, remarking that, contrary to Legendre's opinion, the other imperfections of principle, such as the definition of the straight line, should not be dealt with here, and have no influence whatsoever upon the theory of parallels. (Lobatschewsky, 1866, pp. 87-88)

Lobatschewsky was deeply concerned with Legendre's influential work. In particular, he felt the need to attack Legendre's analytical argument. Indeed, in Lobatschewsky's theory, there do exist relations between the size of a triangle and the sum of its angles (Lobatschewsky, 1837); therefore, it was essential for him to show that this is not absurd.

Given that Lobatschewsky regarded the actual laws of geometry as empirical truths, which cannot be discovered *a priori*, he compared them to physical laws (Daniels, 1975). Observation must show whether there is any natural unit of length in nature:

> Our theory of the parallels establishes among lines and angles some dependence that nobody has been able to show [...] whether it is or is not found in nature. We must at least infer from the astronomical data that all the lines that we are

able to measure, and even the distances between the celestial bodies, are very small compared to the line that plays the role of unit in our geometry. (Lobatschewsky, 1829, p. 22)

Without paying much attention to the details of Legendre's analytic method, Lobatschewsky directly attacked its central assumption about the homogeneity of equations. He stated that these is no reason to suppose that only abstract numbers (ratios of similar magnitudes) may appear in the relations; and he draws a simile between the law of gravitation and the possible relation between angles and lines:

> It is not doubtful that forces create all the rest: motion, speed, time, mass, and even the distances and angles. Everything is intimately bound to forces – a link that we do not understand in its essence. Hence, we have not the right of assuming that, in a relation between quantities of so different natures, only the ratios of these magnitudes appear. If a dependence between the ratios seems admissible, why should not the same hold for the magnitudes themselves? [...]
>
> Ask yourself this, how does distance produce this force? How does it happen that there is a link between two so different things in nature? To be sure, we shall never understand this. However, it is true that forces depend on distances; why should distances not depend on angles? The diversity is similar in the two cases. [...] (Lobatschewsky, 1829, pp. 76-77)

We see that Lobatschewsky repeated Leslie's first argument, and he did not discuss the fundamental problems of the analytical method.

Let us proceed to another interesting case: the work of the Canadian mathematician George Paxton Young (1818-1889), who in 1860 rediscovered non-Euclidian geometry (Halsted, 1894).

In a paper published in 1856, Young first discussed Legendre's work (Young, 1856). He had studied Leslie's attack and Playfair's 1812 defense of Legendre, and remarked that "since that time, the validity of Legendre's reasoning seems to have been admitted by the general consent of mathematician" (Young, 1856, p. 520). In his 1860 article, Young again stated that "mathematicians have – by their silence at least – acquiesced in his [Playfair's] verdict" (Young, 1860, p. 341).

Young produced essentially the same argument presented by Sigma. He noticed that Legendre had assumed that when a quantity is *determined* by several others, it may be *expressed* or computed from these and only these. However, Young did not agree with this assumption.

He also remarked that angles are not numbers, and therefore an angle cannot be found without the intervention of some other angle or angular unit. If, for instance, the angle C of the triangle is determined by its three sides a, b, c, it must be a function of the numerical ratios of these sides, *multiplied by an angle*:

$$C = \text{right angle} \times f(b/a, c/a).$$

Since Young did not see any essential difference between angles and lines, he used the same argument and showed that the side c of the triangle could perhaps be computed from its angles A, B, C, by an equation such as (Young, 1856, p. 522):

$$C = \text{unit of linear measure} \times f(A, B, C).$$

After establishing the possibility of non-Euclidian relations, Young proves in his second paper an essential theorem of non-Euclidean geometry: the proportionality between the area of a triangle and the difference between two right angles and the sum of the angles of this triangle.

8. A HISTORICAL EVALUATION OF THE EARLY USES OF DIMENSIONAL ANALYSIS

What was the final influence of Foncenex's and Legendre's uses of dimensional arguments?

The Turin paper tried to provide an *a priori* proof of a physical law – the principle of force composition. Since it became afterwards clear that it is impossible to provide an *a priori* proof of the basic laws of mechanics, its method was regarded as flawed. Similarly, Legendre tried to provide an *a priori* proof of Euclid's fifth postulate. Since in the long run it was understood that this postulate is arbitrary and can be either true or false, depending on the accepted kind of geometry, Legendre's proof should necessarily have some mistaken assumptions.

It seems that the natural consequence of those attempts should be the *discredit* of dimensional analysis. Although some time after these attempts, a well-founded theory of dimensions way built by Fourier and used by Lord Rayleigh and other authors, *those particular instances* of use of dimensional arguments (in the Turin paper and Legendre's book) were not instrumental in establishing a valid method.

One may recognize a negative outcome of those works in Hermann Laurent's doubts about the very principle of homogeneity, first published in 1870 (Laurent, 1870, pp. 322-326).[18] Although Laurent cited the name of no author, it seems that he had in mind Legendre's work. He discussed the triangle argument and Leslie's criticism, and concluded that the homogeneity of geometrical formulae is not an *a priori* truth. But he added:

> Nevertheless, the equations of Geometry are homogeneous relative to lines, and this is due to the fundamental equations, from which all the others are derived, being themselves homogeneous [...]
>
> In Mechanics, much the same as in Geometry, we cannot establish *a priori* the homogeneity of formulae; nevertheless, these formulae are homogeneous, since the fundamental theorems produce homogeneous relations [...] Hence, the homogeneity exists in the mathematical sciences because this homogeneity has been introduced in the fundamental theorems; where it has not been introduced, it does not exist.

---

[18] The same comments appeared, without any change, in the second and third editions of the book, published in 1878 and 1889.

> So, the equations are not homogeneous relative to angles. (Laurent, 1870, pp. 322-324)

Laurent's basic idea was this: there is no *a priori* reason to accept the homogeneity of equations; but this homogeneity is a hereditary property of equations; and in any field where the basic laws are homogeneous relative to same kind of quantity, all its theorems will also be homogeneous relative to that quantity. This allows us to reach some conclusions about the derived laws in a theory where we do already know that the basic laws are homogeneous, but we can say nothing *a priori* about the homogeneity of the basic laws themselves.

This opinion presented by Laurent was certainly not original, since a similar idea was criticized many years before, by Auguste Comte (1798-1857). He argued that the homogeneity of equations *is not* a hereditary property, since from two homogeneous equations of different degrees – for instance, one relative to lengths and another one relative to areas – we may produce another relation that is not homogeneous, by *adding* the two former equations (Comte, 1843, pp. 36-42). Such an addition is a mathematically correct derivation, since from $A=B$ and $C=D$ we may always derive $A+C=B+D$. In order to forbid the addition of equations of different degrees, it would be necessary to postulate the principle of homogeneity. Hence, Laurent's idea about the hereditary property of homogeneity is only true if we already assume the validity of the principle of homogeneity; since Laurent is trying to show that we do not need this principle, his argument fails.

Comte presented the meaning of the principle of homogeneity relating it to the problem of arbitrariness of units, using Fourier's ideas, which he certainly knew (Comte, 1892, vol. 1, pp. 181-184). Comte's approach was correct, and Laurent's was wrong; but since neither the Turin paper nor Legendre have presented a clear view concerning the principle of homogeneity, Laurent's doubts and criticisms are understandable, as being the most natural reaction to these earlier incorrect attempts to use dimensional analysis.

As a final evaluation of the early history of dimensional analysis, we may state that although Foncenex and Legendre had the priority of using this method, their model was not to be followed at that time. Their motivation was a natural undertaking of their time – the *a priori* proof of fundamental laws in mathematics and physics – but the concept of science underlying their works was soon abandoned. The method and concepts used in their contributions were not clearly understood, and some of their characteristics were in open disagreement with the concept of magnitudes and their relations, at that time. The controversies that followed Legendre's publication have shown that nobody had a very clear idea about the subject, and the scientific community did not reach an agreement on it in the following years: some authors seemed to think that Legendre had been refuted, other believed that he had successfully replied to all criticisms.

We may say that the Turin paper and Legendre's work were, in a sense, ahead of their time – but in a bad sense: something was lacking in their method, something that only became available much time after the earlier uses of dimensional analysis. A correct development of dimensional methods had to wait for an elucidation of the concept of dimensions and of the principle of homogeneity. This was done only in 1822, by Fourier. It was probably for this reason that later authors have progressively forgotten the early history of dimensional analysis, and the opinion was gradually built that everything had begun with Fourier.

ACKNOWLEDGMENT

This work has been supported by the Brazilian National Council for Scientific and Technological Development (Conselho Nacional de Desenvolvimento Científico e Tecnológico – CNPq).

BIBLIOGRAPHIC REFERENCES

BELTRAMI, Eugenio. Un precursore italiano di Legendre e Lobachevsky. *Rendiconti dell'Accademia dei Lincei*, **5**: 441-448, 1889.

BONOLA, Roberto. *Non-euclidean geometry: a critical and historical study of its development.* New York: Dover, 1955.

CARNOT, Lazare. *Géométrie de position.* Paris: J. B. M. Duprat, an XI [1803].

COMTE, Auguste. *Cours de philosophie positive.* 6 vols. Paris: Société Positiviste, 1892-94.

COMTE, Auguste. *Traité élémentaire de géométrie analytique a deux et a trois dimensions.* Paris: Carilian-Goeury et Vor Dalmont, 1843.

DANIELS, Norman. Lobachevsky, some anticipations of later views on the relation betweengeometry and physics. *Isis*, **66**: 75-85, 1975.

DELAMBRE, Jean-Baptiste. *Discours sur les progrès des sciences, lettres et arts, depuis MDCCLXXXIX jusqu'à ce jour, ou, Compte rendu par l'Institut de France à S.M. l'empereur et roi.* En Hollande: chez Immerzeel et Compagnie, 1809.

DELBŒUF, Joseph. L'ancienne et les nouvelles géométries. IV. Les axiomes et les postulats de la géométrie de l'espace homogène. *Revue Philosophique de la France et de l'Étranger*, **39**: 345-371, 1895.

DIS-IOTA. Further remarks on the theory of parallel lines. *The Philosophical Magazine*, **63** (312): 246-252, 1824.

DIS-IOTA. On the application of algebraic functions to prove the properties of parallel lines. *The Philosophical Magazine*, **63** (311): 161-167, 1824.

EINSTEIN, Albert. *The meaning of relativity.* 5th edition. London: Methuen, 1951.

ENGEL, Friedrich; STÄCKEL, Paul (eds.). *Die Theorie der Parallellinien von Euklid bis auf Gauss: eine*

Urkundensammlung zur Vorgeschichte der nichteuklidischen Geometrie. Leipzig: Teubner, 1895.

FONCENEX, François Daviet de. Sur les principes fondamentaux de la mechanique. *Mélanges de Philosophie et Matématique de la Société Royale de Turin*, **2**: 299-322, 1761.

GONSETH, Ferdinand. *La géometrie et le problème de l'espace: le problème de l'espace*. Paris: Dunod, 1955.

HALSTED, George B. *Girolamo Saccheri's Euclides vindicates*. Chicago: Open Court, 1920.

HALSTED, George Bruce. The non-Euclidean geometry inevitable. *The Monist*, **4** (4): 483-493, 1894.

HEATH, Thomas Little. *The thirteen books of Euclid's Elements*. 2 vols. New York: Dover, 1956.

HILL, Micaiah John Muller. On the teaching of mathematics. Presidential Address to the Mathematical Association, 1927. *The Mathematical Gazette*, **13** (187): 296-312, 1927.

IVORY, James. On the theory of parallel lines in geometry. *The Philosophical Magazine*, **59** (287): 161-167, 1822.

LAMBERT, Johannes Heinrich. Die Theorie der Parallellinien. *Leipziger Magazin für reine und angewandte Mathematik*, **3**: 325-358, 1786.

LAPLACE, Pierre-Simon. *Exposition du système du monde*. 6th edition. Paris: Bachelier, 1835. Reprinted in: Œuvres Complètes de Laplace, vol. 6. Paris: Gauthier-Villars, 1884.

LAURENT, Hermann. *Traité de mécanique rationelle, à l'usage des candidats à la licence et à l'agrégation*. 2 vols. Paris: Gauthier Villars, 1878.

LEGENDRE, Adrien-Marie. *Éléments de géométrie*. Avec des notes. Paris: Firmin Didot, 1794.

LEGENDRE, Adrien-Marie. *Éléments de géométrie*. Avec des notes. 3rd edition. Paris: Firmin Didot, 1800.

LEGENDRE, Adrien-Marie. *Éléments de géométrie*. Avec des notes. 12th edition. Paris: Firmin Didot, 1823.

LEGENDRE, Adrien-Marie. *Elements of geometry and trigonometry*. Trans. David Brewster. Edinburgh: Oliver & Boyd, 1824.

LEGENDRE, Adrien-Marie. Réflexions sur différentes manières de démontrer la théorie des parallèles, ou le théorème sur la somme des trois angles du triangle. *Mémoires de l'Académie des Sciences*, **12**: 367-411, 1833.

LEMAÎTRE, Georges. Cosmological application of relativity. *Reviews of Modern Physics*, **21** (3): 357-366, 1949.

LESLIE, John. *Elements of geometry, geometrical analysis, and plane trigonometry*. Edinburgh: John Ballantyne, 1811.

LESLIE, John. *Elements of geometry, geometrical analysis, and plane trigonometry*. 2nd edition. Edinburgh: Archibald Constable, 1817.

LOBATSCHEWSKY, Nikolai Ivanovitch. *Études géométriques sur la théorie des parallèles*. Translated by Jules Hoüel. *Mémoires de la Société des Sciences Physiques et Naturelles de Bordeaux*, **4**: 83-182, 1866.

LOBATSCHEWSKY, Nikolai Ivanovitch. *Géométrie imaginaire. Journal für die reine und angewandte Mathematik*, **17**: 295-320, 1837.

LOBATSCHEWSKY, Nikolai Ivanovitch. *Zwei geometrische Abhandlungen*. Leipzig: B. G. Teubner, 1898-99. Séries: Urkunden zur geschichte der nichteuklidischen geometrie, hrsg. von F. Engel und P. Stäckel

LORIA, Gino. *Il passato e il presente delle principali teorie geometriche: storia e bibliografia*. Padova: CEDAM, 1931.

MAURICE, Frédéric. Sur l'application de l'algorithme des fonctions a la démonstration des propositions fondamentales de la géométrie. *Bibliothèque Universelle des Sciences, Belles-Lettres, et Arts*, **12** (2): 85-98, 1819.

MEIKLE, Henry. On the theory of parallel lines in geometry: To the editors of the Philosophical Magazine and Journal. *The Philosophical Magazine*, **60** (296): 423-426, 1822.

PIONCHON, Joseph. *Théorie des mesures: introduction a l'étude des systèmes de mesures usités en physique.* Paris: Gauthier-Villars, 1891.

PLAYFAIR, John. Discours sur le progrès des sciences, lettres et arts, depuis 1789 jusqu'à ce jour (1805). *The Edinburgh Review or Critical Journal*, **15** (29): 1-24, 1810. This review was not signed, but its author was easily identified by Leslie.

PLAYFAIR, John. Review of John Leslie's Elements of geometry, geometrical analysis & plane trigonometry. *Edinburgh Review*, **20**: 79-100, 1812.

RAYLEIGH, John William Strutt, Lord. *The theory of sound.* 2 vols. New York: Dover, 1945.

RINDLER, Wolfgang. Relativistic cosmology. *Physics Today*, **20** (11): 23-31, 1967.

ROBERTSON, Howard Percy. Relativistic cosmology. *Reviews of Modern Physics*, 5 (1): 62-90, 1933.

SACCHERI, Girolamo. *Euclides ab omni naevo vindicatus: Sive conatus geometricus quo stabiliuntur prima ipsa universae geometriae principia.* Mediolani: Pauli Antonii Montani, 1733

SANDAGE, Allan R. Cosmology: A search for two numbers. *Physics Today*, **23** (2): 34-41, 1970.

SIGMA. On the use of analysis in investigating the elementary relations of figure and forces. *The Philosophical Magazine*, **65**: 195-200, 1825.

SIGMA. On the use of functional equations in the elementary investigations of geometry. *The Philosophical Magazine*, **65**: 101-105, 1825.

SIGMA. Outline of general methods for the development of certain branches of analysis. *The Philosophical Magazine*, **65**: 344-354, 1825.

SJÖSTEDT, Carl Erik. *Le axiome de paralleles de Euclide a Hilbert: un probleme cardinal in le evolution del geometrie.* Excerptes in facsimile ex le principal ovres original e traduction in le lingue international auxiliari "Interlingue". Uppsala: Interlingue-Fundation, 1968.

SMITH, David Eugene. *A source book in mathematics*. New York, McGraw-Hill, 1929.

SOMMERVILLE, Duncan M. Y. *Bibliography of non-Euclidean geometry, including the theory of parallels, the foundations of geometry, and space of n dimensions*. London: Harrison and Son, 1911.

WALLIS, John. *Operum mathematicorum*. 2 vols. Oxoniæ: E Theatro Sheldoniano, 1693.

WALSH, John. Observations on the twelfth book of Euclid. *The Philosophical Magazine*, **64** (317): 181-184, 1824.

WALSH, John. On parallel straight lines. *The Philosophical Magazine*, **63** (312): 271-273, 1824.

WOLFE, Harold Eicholz. *Introduction to non-Euclidian geometry*. New York: Holt, Rinehart and Winston, 1945.

YOUNG, George Paxton. An examination of Legendre's proof of the properties of parallel lines. *Canadian Journal of Industry, Science, and Art* [new series] **10**: 519-522, 1856.

YOUNG, George Paxton. The relation which can be proved to subsist between the area of a plane triangle and the sum of the angles, on the hypothesis that Euclid's 12th axiom is false. *Canadian Journal of Industry, Science and Art*, **5**: 341-356, 1860.

EXPERIMENTAL STUDIES ON MASS AND GRAVITATION IN THE EARLY TWENTIETH CENTURY: THE SEARCH FOR NON-NEWTONIAN EFFECTS

Roberto de Andrade Martins

Abstract: From the late nineteenth century to the early twentieth century, before the development and acceptance of the theory of general relativity, much attention was devoted to the experimental search of some hypothetical non-Newtonian gravitational effects. The influences that were investigated included eventual violations of the principle of equivalence between gravitational and inertial masses; the influence of radioactivity on gravitation; the hypothetical absorption of gravity by matter; the influence of temperature on gravitation; and the influence of chemical reactions upon weight. This paper presents an overview of those experimental investigations, which were mostly forgotten.

Keywords: gravitation; mass; non-Newtonian effects; experiments on gravitation

1. INTRODUCTION

From Isaac Newton onwards, it has been generally accepted that the gravitational attraction between two bodies depends only on their masses, distance, and geometrical factors (size and shape of the bodies). However, the young Newton himself suspected that the gravitational force could be influenced by several other factors, and he wrote in one of his notebooks

MARTINS, Roberto de Andrade. *Studies in History and Philosophy of Science I*. Extrema: Quamcumque Editum, 2021.

(*Quæstiones quædam philosophicæ*, *circa* 1664) a list of experiments that should be tried to check whether some effects did exist:

> Try whither the weight of a body may be altered by heate or cold, by dilatation or condensition, beating, poudering, transfering to severall places or severall heights or placing a hot or heavy body over it or under it or by magnetisme whither leade or its dust spread abroade, whither a plate flat ways or edg ways in heaviest, whither the rays of gravity may bee stopped by refecting or refracting them, if so a perpetuall motion may bee made one of these two ways. (Newton, MS Add. 3996, Cambridge University Library, fol. 121v; McGuire & Tamny, 1983, p. 430)

In his student days, when Newton began to develop his new cosmological outlook, he pondered about the cause of gravity. The notes he took at this time allows us to get an insight into his early thoughts. Following – as it seems – the ideas of Kenelm Digby, Newton speculated about a downwards flow of matter that would press the bodies towards the Earth. According to this hypothesis, there would be a steady flow of particles from the heavenly space towards the centre of the Earth, and from the Earth to the space. Those particles would act upon bodies, through impact, both at their surfaces and through their pores. However, if this were the cause of gravity – so Newton thought – there would be several observable consequences (McGuire & Tamny, 1983, pp. 279-282). The effect of gravity upon a swiftly falling body would be smaller than upon the same body at rest – because the impacts would have a smaller effect. The gravitational push could also, in principle, be altered by a change of the condensation of the body. Hypothetically, the flux of particles could also be altered by heat, by magnetism or other effects.

We don't know whether Newton made or did not make all those tests. However, in his mature writings – more specifically, in the *Principia* – we find no suggestion similar to those that were registered in his Notebook. If gravitation were produced by any kind of penetrating particles, one could expect that the gravitational force would not be *exactly* proportional to the amount of matter of very large bodies – such as the planets and the Sun. However, Newton noticed that this proporcionality seemed to hold. This was a strong argument against any "ether stream" hypothesis.

In the century following the publication of the *Principia*, as Newton's gravitational theory became widely accepted, it seemed to most people that ho hypothesis was needed to explain gravitational forces. The concept of forces at a distance ("attraction", as it was simply called) did not lead to any search for strange effects.

However, in the 19th century, the situation slowly changed. The study of electricity and magnetism led to new ideas – and, although there were approaches such as Weber's, that used distant action, Maxwell's successful theory introduced an intervening aether. The discovery of energy conservation was regarded by several people – including Faraday – as a sign of an underlying unity of all physical forces. Accordingly, there could be a relation between electromagnetism and gravitation, in two relevant senses: there could be a similarity between their modes of action; and there could be gravito-electromagnetic phaenomena (as there were electro-magnetic effects). Under the influence of those developments, there arose theoretical speculations for a *mechanism* of gravitation and a parallel search for non-Newtonian effects.

Towards the end of the 19th century, there was an impressive rise of speculations concerning the nature and mode of action of gravitation. Most of those speculations were mere explanatory hypotheses: they just tried to explain known facts, without suggesting any new phaenomena. However, the bare assumption of any kind of mediated gravitational action opened the way to considerations about the possibility of influencing the gravitational force. Towards the end of the 19th century, a new field of research was developing: the experimental study of gravitation and the search for non-Newtonian effects (Woodward, 1972).

From the last decade of the 19th century to the decade of 1920, there was an intensive exploratory experimental research on the properties of mass and gravitation. Most of this work has been forgotten, since no new regular effects were observed. Besides that, the rise and development of the theory of general relativity – which did not predict any of those effects – led to a fast neglect of that exploratory activity. However, several interesting investigations have been developed by a lot of competent researchers. The history of those researches deserves a detailed study.

The main episodes that I have been able to detect are shortly described below[1].

2. CONSERVATION OF MASS IN CHEMICAL REACTIONS

During the 19th century, there was a general acceptance of the law of mass conservation in chemical reactions. However, in 1891 Damian Kreichgauer observed small variations of the weight of a closed system where a chemical reaction occurred. The observed changes were of about 1/20,000,000. Shortly after this publication, Hans Landolt began a series of tests of the law of mass conservation (see Martins, 1993; Martins, 2019). His first results were presented in 1893. He observed weight

---

[1] Most of this paper was contained in a research project written in 1994. Its Introduction was added in 1995, when I presented a talk on this subject at the Laboratoire de Gravitation et Cosmologie Relativistes, Université Pierre et Marie Curie (Paris, France). The paper has not been previously published. Only a shorter version, in Portugueses, appeared a few years later (Martins, 1997). When the paper was written, I had already published accounts of the episodes described in sections 2 and 4, in Portuguese. I have not expanded or updated the content of the paper; I have only made slight changes. I also added the bibliographic references of some of my later publications on those subjects.

changes as large as 1/1,000,000 in chemical reactions produced inside hermetically closed glass tubes.

Landolt's work was reproduced in several journals. Several researchers tried to reproduce his experiments, with different results. Fernando Sanford and Lilian Ray (1897, 1898) and Antonino Lo Surdo (1904, 1906) observed no change of weight. Adolf Heydweiller (1900, 1901) detected relevant weight reductions in several experiments.

All those experiments measured the weight changes (that is, variation of passive gravitational mass) of the system containing the chemical substances. Lord Rayleigh (1901) suggested an important question: are there changes of inertia related to those weight changes? This question was investigated by Joly (1903), who used a delicate torsion balance. He detected a regular effect, but much smaller than was expected from the previous experiments.

Landolt resumed his experiments (1906). He was able to build a balance that could detect mass changes of 1/10,000,000. With an improved experimental technique, some reactions now exhibited no significative mass change.

At this time, the theory of relativity was being developed and there appeared the famous mass-energy relation $E=m.c^2$. Max Planck (1907) discussed the possibility that the observed effect could be due to energy exchanges, but he computed that the relativistic effect would be much smaller than that measured by Landolt.

Landolt (1908) and Constantin Zenghelis (1909) discussed the possibility that the observed mass variation could be due to the passage of small quantities of matter through the glass of the vessels. Landolt also studied the effect of small temperature and humidity variations in the experiments. After the elimination of several sources of systematic error, the measured mass variations were gradually reduced – but not eliminated.

After Landolt's death, in 1910, all those studies have been quickly forgotten. It seems that there was only one single new experiment, by John Manley (1913). He reproduced those of

Landolt's experiments that had produced the smallest mass variations and he observed no change of weight greater than 1/100,000,000.

Several years later, Roland von Eötvös and his group (1922) published some measurements of the ratio of inertial and gravitational masses for some of the chemical substances used by Landolt and Heydweiller. No anomalies were observed. The conclusion was that, within the limits of the experimental error (1/100,000,000) there was no change of the ratio of inertial to gravitational mass in chemical reactions.

3. INFLUENCE OF TEMPERATURE ON MASS AND GRAVITATION

It is well known that during the 18th century there was much speculation about the role of heat (caloric, flogiston) in chemical reactions and about its weight. The chemical aspects of this subject were elucidated by the works of Lavoisier, Cavendish, etc. On the physical side, the subject was investigated by Count Rumford (1799). He established that the freezing of water produces no change of its weight, within the experimental errors of 1/1,000,000.

During the 19th century, Augustin Fresnel (1825) and other researchers have observed an effect of repulsion due to the heating of bodies in rarefied air. William Crookes (1874) observed the same kind of effect. After a series of experiments (Crookes 1875-1879), he concluded that the repulsion was due to the residual air (radiometer effect) – it was not a direct repulsion produced by hot bodies.

Some anomalies were also observed in the measurement of the gravitational constant. In 1883, William Hicks used the date of Baily's 1842 research on gravitation and noticed a correlation between temperature and the observed attraction in a Cavendish balance. There seemed to be a steady increase of attraction with temperature – a variation of 7,8% for a 100° F.

Hick's paper motivated John Henry Poynting and Percy Phillips (1905) to develop measurements of weight of bodies at different temperatures. They have observed no regular effect. Hicks himself also developed an apparatus to measure the influence of temperature on weight. The measurements were made by Leonard Southerns (1907) who was unable to find any regular effect, too.

Those results were not incompatible with Hicks' observations, because only the temperature of the attracted test body was changed in the experiments of Poynting and Phillips, and of Southerns. It was suggested that they were inconclusive, since the temperature of the Earth could not be changed.

In 1916, Philip Shaw published the result of a long series of measurements of gravitational attraction, using a torsion balance. He detected a regular increase of attraction with temperature, corresponding to 0,12% for a variation of 100° C. Shaw's work was presented to the Royal Society by Boys – the greatest authority in gravitational measurements of that time.

There was an immediate reaction of the community. Several letters were published in *Nature*, discussing theoretical aspects of the relation between temperature and gravitation. No other researcher tried to repeat Shaw's experiments. But Shaw himself, with the help of Cecil Hayes (1917), improved his measurements and the observed effect was about 10% greater than the previous results. But Shaw was not satisfied with the experiment. After improving the control of the position of the attracting bodies, the effect vanished, as described in the last paper by Shaw and Norman Davy (1923). The final conclusion was that any thermal effect must be smaller than $2 \times 10^{-6}/°C$.

4. ABSORPTION OF GRAVITY BY MATTER

At the end of the 19th century several authors speculated about possible effects of intervening matter between two attracting bodies (see Martins, 1993; Martins, 1999; Martins, 2002a; Martins, 2002b; Martins, 2004). John Henry Poynting (1900) suggested that matter could affect the gravitational force between two bodies either as occurs in electromagnetism, or that the "gravitational rays" could be absorbed as light or radiation.

In 1897, Louis Austin and Charles Thwing studied the effect of interposing screens of different materials between the attracting bodies in a torsion balance. The attraction of the screen itself was compensated by the experimental arrangement. Any temperature changes were avoided, together with several other experimental precautions.

Austin and Thwing tested the effect of different groups of substances, such as lead and mercury (because of their high density) or water, alcohol and glycerin (for their high dielectric constant). Only in the case of iron screens there was a measured effect greater than the expected errors (of about 1/500). The conclusion was that there was no effect similar to "gravitational permeability" and that the iron effect was spurious (due to magnetic – not gravitational – forces).

Fritz Laager, in 1904, published the results of similar experiments with null results. He used spherical shells around the test bodies, in order to avoid the attraction of the interposed matter. In 1908, Theodor Erismann also published the result of investigations of the same kind, with no effect greater than the experimental errors (of about 1/1,000).

The next year, Crémieu (1905) described a strange effect. He suspended small olive oil drops in a solution of water and alcohol. The density of the solution was equal to the density of the drops and they remained in equilibrium in this solution. However, after some time, the drops began to approach to one another. There was no classical explanation for this phenomenon: the approach could not be ascribed to gravitation, capillary or hydrodynamic effects.

These observations led Victor Crémieu to measure gravitational forces inside a liquid, using a Cavendish balance (Crémieu, 1905-1907). There seemed to be an increase of the gravitational attraction, in water, of about 7%. But after a long series of investigations, Crémieu found some unexpected problems with the experiment. He afterward concluded that it is impossible to compare the experiments inside water to experiments in air because of systematic errors amounting up to 10% (Crémieu 1909-1917).

In 1909, Hugo von Seelinger suggested that the attraction between the Moon and the Sun could decrease during eclipses, due to the absorption of gravity by the Earth. Kurt Bottlinger (1912) developed the theory of this effect and compared the computed effect to observed irregularities in the longitude of the Moon. There was a nice agreement, and Bottlinger computed that the maximum decrease of gravitational attraction between the Moon and the Sun was of 1/60,000.

Bottlinger's results were criticized by Willem De Sitter (1912, 1913). Using slightly different auxiliary hypotheses, De Sitter obtained results that were similar to those of Bottlinger for small time periods, but widely different for large periods. However, latter studies by Bottlinger (1914) showed even better agreement between theory and observation than his previous results.

The Italian physicist Quirino Majorana brought the problem back to laboratory. First, from astronomical data, he was able to limit the value of the gravitational absorption of gravity to 7.65×10^{-12} cm^2/g (Majorana 1919). He then devised very delicate experiments to detect this small effect. His measurements, using mercury and lead, led to absorption constants of 6.7×10^{-12} and 2.5×10^{-12} cm^2/g, respectively (Majorana 1919-1921).

Majorana's experiments were discussed by Henry Norris Russell (1921). He argued that the astronomical consequences of the effect would be too high to escape notice. However, Eddington (1922) showed that Russell's arguments were not conclusive. The main problem of the concept of gravitational absorption, according to Arthur Eddington, would be the possibility of a gravitational perpetual motion.

No other scientist repeated Majorana's experiments. Roland von Eötvös *et al.* (1922) tried to detect gravitational absorption using a torsion balance, with null results. However, even after the improvement of his instruments, Majorana still obtained gravity absorptions greater than 10^{-12} in 1930.

5. THE MASS OF RADIOACTIVE SUBSTANCES

The discovery of radioactivity led to the suspicion that several basic physical laws should be changed (Lodge 1912). The continuous heat developed by radioactive substances (first measured in 1903 by Pierre Curie and Albert Laborde), was very difficult to explain. One of the several suggested explanations was that radioactive bodies obtain their energy from the gravitational field – therefore their weight should exhibit some kind of anomaly.

Adolf Heydweiller (1902) was the first to test the constancy of weigh of radioactive substances. He suspended a glass tube with 5 g of strongly radioactive substance to a balance and observed a decrease of its weight (-0.02 mg per day). After a few weeks the weight reduction amounted to 0.5 mg.

Heydweiller tried to explain this decrease. The gravitational potential energy of 0.02 mg corresponds to 1.2×10^7 erg. The heat generated by the radioactive material amounted to about 10^7 erg per day. Heydweiller concluded that the radioactive body transformed its gravitational energy in heat.

This experiment was repeated by Ernst Dorn (1903). He used a small piece of radioactive material (just 30 mg) and observed a weight reduction of 0.001 mg after three months. Since his sample was ten times more radioactive than Heydweiller's, it should have exhibited a much higher weight variation, according to the energy transformation hypothesis.

Robert Geigel (1903), Carl Forch (1903) and Walter Kaufmann (1903) tested the effect of placing a radioactive body between a test body and the Earth. Geigel observed a reduction of the weight of the test body and ascribed the effect to the absorption of "gravitational rays". Foch observed no effect. Kaufmann observed effects similar to those measured by Geigel, but explained the weight reduction by convection currents produced by the heat generated by the radioactive substance.

A few years later, Georges Sagnac (1906) tried to detect any difference between the inertial masses of equal weights of barium and radium compounds. His experiment could observe only large differences (of about 1%). No irregularity was noticed.

Joseph John Thomson (1909) and Leonard Southerns (1911) compared the ratios of gravitational to inertial mass of radioactive bodies and their decay products (such as lead). Thomson's pendulum experiments established an upper limit of 1/2,000 to any variation of this ratio. Southerns, using Bessel's method, obtained an upper limit of 1/200,000.

Eötvös also tried to detect any change of the ratio of gravitational to inertial mass of radioactive substances, using a torsion balance. The experimental setup could observe changes of 1/2,000,000. No effect was detected (Eötvös *et al.*, 1922). Zeeman (1918) has also used a torsion balance to test the same effect, with a similar result.

6. VIOLATION OF THE PRINCIPLE OF EQUIVALENCE

Several of the experiments on mass and weight of the early 20th century were motivated by speculations concerning the existence of gravitational waves or rays. In 1911, Charles Brush suggested that gravitation is produced by electromagnetic waves of very short wavelength. Those waves should be able to pass through matter with little absorption, except in the case of a few substances. Diamagnetic bodies such as bismuth – should, however, exhibit high absorption and anomalous gravitational effects (Brush, 1914).

Brush compared the gravitational attractions of zinc and bismuth, using a Cavendish balance. According to his report (Brush 1921-1922), the attraction produced by bismuth was only 72% of the attraction produced by an equivalent weight of zinc. He also made pendulum experiments and measured a difference of about 1/50,000 between the ratio of inertial and gravitational mass for those substances.

Harold Potter (1922) and Harold Wilson (1922) tested the proportionality between inertial and gravitational mass for bismuth and other substances and detected no anomaly. Potter's experiments were not sensitive enough, but Wilson, using an Eötvös balance, reached a sensitivity of 1/1,000,0000.

Potter produced a series of further experiments (Potter, 1923, 1927) to test the proportionality between inertial and gravitational mass of several substances. The main motivation of those investigations was to test whether the nuclear structure of the chemical elements could affect their mass ratio. The conjecture that hydrogen could have an anomalous behaviour had been suggested by Owen Richardson (1910, 1922). However, Potter was unable to detect any effect.

Notwithstanding the negative results of other researchers, Brush went on with his studies. In a series of articles (Brush, 1923-1928) he presented further anomalous results. In a free fall experiment, he measured a difference between the accelerations of lead and aluminum of about 1/10,000. The crystalline state of the material also seemed to affect the results.

Brush speculated that gravitational attraction should be produced by gravitational waves and that some substances could transform the energy of those waves and exhibit a spontaneous heating. Using a calorimeter (Brush, 1926) he indeed detected that several non-radioactive substances maintained a higher temperature than their environment. This was the case for several complex silicates. He also described that those substances had a smaller gravitational acceleration.

Brush's work has been largely ignored by the scientific community both at the time of his original publications and afterwards.

7. OTHER INVESTIGATIONS

Several other researches were developed during the period from the last decade of the 19th century to about 1920. Arthur

Mackenzie (1895), John Henry Poynting and Peter Gray (1899), Pter Zeeman (1918) and Paul Heyl (1924) looked for the existence of an anisotropy of the gravitational attraction of crystals. No regular effect was detected.

A secular increase of the gravitational acceleration at Paris was claimed by Édouard Caspari in 1895. He used the results of measurements from 1794 to 1890 and obtained a variation of about 1/8,000 in one century. Karl Koch, in 1904, described the variation of the gravitational acceleration at Stuttgart. He measured an increase of about 1/300,000 in four years – a value consistent with Caspari's results.

Periodical variations of the gravitational field of the Earth have been claimed by several authors. Gerald Drossbach (1895) observed a daily variation of gravity of about 1/1,200. Ralph Hartsough (1922) observed a small effect that was ascribed to lunar tides. Leopold Courvoisier (1927) claimed that there was a daily variation of gravity due to the Lorentz contraction of the Earth (Martins, 2011). Rudolf Tomaschek and Walter Schaffernicht (1932) looked for short period variations of the gravitational field of the Earth. They were able to measure the tidal effects of Moon and Sun and found no anomalous result. A strange effect was however claimed by Russel Goudey (1922). He described an annual variation of the gravitational acceleration of about 1/1,000,000, detected by the comparison between astronomical observations and pendulum measurements.

A few experiments tried to find any influence of electricity or magnetism upon gravitation. In 1909, Southerns detected a difference between the weight of bodies with positive and negative electrical charge. L. Simons (1922) made experiments very similar to Southerns' and obtained no anomalous result. Paul Agnew and William Bishop (1912) measured the weight of a capacitor and found no difference between the charged and uncharged states. Francis Nipher (1917) used a shielded torsion balance and described a strong influence of the electric potential of the system upon gravitational attraction.

Louis Bauer (1907, 1909) noticed that the weight of a magnet varied with its position, even in a region where the magnetic field of the Earth was uniform. He also measured a weight difference between unmagnetized and magnetized iron. Lloyd (1909) repeated Bauer's experiments and noticed no anomaly.

Several authors suggested, in the 19th century, that the gravitational forces could be influenced by the velocity of the attracting bodies. In 1896, the brothers Benedict and Immanuel Friedländer tested the influence of a gyroscope upon a torsion balance. They found a small effect. August Föppl (1905) and Victor Crémieu (1917) also tried to measure the influence of a gyroscope upon a common balance and upon a torsion balance. They noticed no effect.

8. GENERAL COMMENTS

This short overview shows that from the end of the 19th century to the 1920's there has been intense experimental research on gravitation and the search for several non-Newtonian effects. It seems that those episodes have been forgotten by both physicists and historians of science[2].

This kind of experimental effort subsides and seems to disappear after 1930. It is likely that the confirmation of general relativity was the reason of this historical change. The conjectured anomalous effects that were investigated were not compatible with the theory of general relativity (or, if compatible, the predicted effects were several orders of magnitude smaller than the available experimental sensibility). As confidence in the theory of general relativity grew, the search for such effects was regarded as worthless effort[3].

----

[2] Some of the above described experiments have been cited in recent review papers on gravitation (see, for instance, Cook 1988). But it seems that no systematic and critical study of those episodes has ever been produced.

[3] This is a historical hypothesis that should be investigated. In the case of Landolt's studies on mass variation in chemical reactions, there was

It seems to me that this set of researches, together with their broader scientific context, deserves a careful study. Maybe some of the scientists that devoted their time to those experiments were just cranks[4]. But some of them were highly respected scientists. It seems that the search for non-Newtonian gravitational effects was probably regarded as a respectable, relevant part of scientific research during about three decades.

BIBLIOGRAPHIC REFERENCES

AGNEW, Paul Gough; BISHOP, W. C. An attempt to detect possible changes in weight or momentum effects on charging a condenser. *Physical Review*, **35**: 470-476, 1912.

AUSTIN, Louis W.; THWING, Charles B. An experimental research on gravitational permeability. *Physical Review*, **5**: 294-300, 1897.

BAILY, Francis. An account of some experiments with the torsion rod, for determining the mean density of the Earth. *Memoirs of the Royal Astronomical Society* **14**: 1-120, i-cclviii, 1843; *Monthly Notices of the Royal Astronomical Society* **5**: 188-196, 197-206, 1839-43; *Annales de Chimie* **5**: 338-353, 1842; *Bibliothèque Universelle* **43**: 177-181, 1843; *London, Edinburgh and Dublin Philosophical Magazine* **21**: 111-21, 1842; *Annalen der Physik und Chemie* **58**: 453-467, 1842.

BAUER, Louis A. Results of careful weightings of a magnet in various magnetic fields. *Physical Review* **25**: 498-499, 1907

BAUER, Louis A. Is the Earth's action on a magnet only a couple? *Terrestrial Magnetism and Atmosferic Electricity* **13**: 25-35, 1908.

---

a clear influence of the special theory of relativity (the mass-energy relation) in the dismissal of the anomalous effect.

[4] Brush is a likely candidate to this group.

BAUER, Louis A. On the apparent alterations of mass disclosed by weighings of magnets. *Terrestrial Magnetism and Atmosferic Electricity* **14**: 72-76, 1909.

BOTTLINGER, Kurt Felix Ernst. Die Erklärung der empirischen Glieder der Mondbewegung durch die Annahme einer Extinktion der Gravitation im Erdinnern. *Astronomische Nachrichten* **191**: col. 147-150, 1912.

BOTTLINGER, Kurt Felix Ernst. *Die Gravitationstheorie und die Bewegung des Mondes*. Freiburg: C. Trömer, 1912.

BOTTLINGER, Kurt Felix Ernst. Zur Frage der Absorption der Gravitation. *Sitzungsberichte der Königl Bayerische Akademie der Wissenschaften zu München. Mathematische-physikalische Klasse* **44**: 223-239, 1914.

BRUSH, Charles F. A kinetic theory of gravitation. *Science* **33**: 381-386, 1911 (a).

BRUSH, Charles F. A kinetic theory of gravitation. *Nature* **86**: 130, 1911 (b).

BRUSH, Charles F. Discussion of "A kinetic theory of gravitation". *Proceedings of the American Philosophical Society* **53**: 118-128, 1914.

BRUSH, Charles F. Discussion of a kinetic theory of gravitation II, and some new experiments in gravitation[5]. *Physical Review* **18**: 125-126, 1921.

BRUSH, Charles F. Discussion of a kinetic theory of gravitation, II; and some new experiments in gravitation. *Proceedings of the American Philosophical Society* **60**: 43-61, 1921; **61**: 167-183, 1922; **62**: 75-89, 1923; **63**: 57-61, 1924; **64**: 36-50, 1925.

BRUSH, Charles F. Discussion of a kinetic theory of gravitation III. Some experimental evidence supporting theory; continual generation of heat in some igneous rocks and minerals. Relation of this to the internal heat of the Earth

---

[5] This paper is a summary of the first paper of equal title published in the *Proceedings of the American Philosophical Society*.

and presumably of the Sun. *Proceedings of the American Philosophical Society* **65**: 207-231, 1926.

BRUSH, Charles F. Discussion of a kinetic theory of gravitation IV: correlation of continual generation of heat in some substances, and impairment of their gravitational acceleration.. *Proceedings of the American Philosophical Society* **67**: 105-117, 1928 (a).

BRUSH, Charles F. Discussion of a kinetic theory of gravitation IV: correlation of continual generation of heat in some substances, and impairment of their gravitational acceleration[6]. *Physical Review* **31**: 1113, 1928 (b).

BRUSH, Charles F. Some experimental evidence supporting the kinetic theory of gravitation. *Journal of the Franklin Institute* **206**: 143-149, 1928 (c).

COOK, Alan. Experiments on gravitation. *Reports of Progress in Physics* **51**: 707-757, 1988.

COURVOISIER, Leopold. Bestimmungsversuche der Erdbewegung relativ zum Lichtather II. *Astronomische Nachrichten*, 230: 425-432, 1927.

CRÉMIEU, Victor. Attraction observée entre gouttes liquides suspendues dans un liquide de même densité. *Comptes Rendus des Séances de l'Académie des Sciences de Paris* **140**: 80-3, 1905 (a)

CRÉMIEU, Victor. Recherches sur la gravitation. *Comptes Rendus des Séances de l'Académie des Sciences de Paris* **141**: 653-6, 713-5, 1905 (b).

CRÉMIEU, Victor. Recherches expérimentales sur la gravitation[7]. *Journal de Physique Théorique et Appliquée* [4] **5**: 25-39, 1906 (a).

---

[6] This is a short abstract of the paper of equal title published in the *Proceedings of the American Philosophical Society*.

[7] This paper is a verbatim reproduction of the communication to the *Société Française de Physique* (1905 c).

CRÉMIEU, Victor. Recherches sur la gravitation. *Comptes Rendus des Séances de l'Académie des Sciences de Paris* **143**: 887-9, 1906 (b).

CRÉMIEU, Victor. Recherches comparées sur les forces de gravitation dans les gaz et les liquides[8]. *Journal de Physique Théorique et Appliquée* [4] **6**: 284-98, 1907.

CRÉMIEU, Victor. Emploi de la balance de torsion comme sismographe. *Comptes Rendus des Séances de l'Académie des Sciences de Paris* **148**: 1161-3, 1909 (a).

CRÉMIEU, Victor. Détermination nouvelle de la constante newtonienne. *Comptes Rendus des Séances de l'Académie des Sciences de Paris* **149**: 700-2, 1909 (b).

CRÉMIEU, Victor. Sur une erreur systématique qui limite la précision de l'éxperience de Cavendish. Méthode nouvelle pour l'étude de la gravitation. *Comptes Rendus des Séances de l'Académie des Sciences de Paris* **150**: 863-6, 1910.

CRÉMIEU, Victor. Effects de la flexion aux points d'attache du fil d'une balance de torsion. *Comptes Rendus des Séances de l'Académie des Sciences de Paris* **156**: 617-20, 1913 (a).

CRÉMIEU, Victor. Séismographes donnant directement les trois composantes d'un séisme et les variations lentes de la verticale. *Comptes Rendus des Séances de l'Académie des Sciences de Paris* **156**: 832-5, 1913 (b).

CRÉMIEU, Victor. Recherches expérimentales sur la gravitation. *Comptes Rendus des Séances de l'Académie des Sciences de Paris* **165**: 586-9, 688, 1917 (a).

CRÉMIEU, Victor. Nouvelles recherches expérimentales sur la gravitation. *Comptes Rendus Hebdomadaires des Séances de l'Académie des Sciences de Paris* **165**: 670-2, 1917 (b).

CROOKES, William. On the action of heat on gravitating masses. *Proceedings of the Royal Society of London* **22**: 37-41, 1873-74.

---

[8] This paper is a reproduction of the communication to the *Société Française de Physique* (1907 a).

CROOKES, William. On attraction and repulsion resulting from radiation. *Proceedings of the Royal Society of London* **23**: 373-8, 1874-75.

CROOKES, William. On attraction and repulsion resulting from radiation. *Philosophical Transactions of the Royal Society of London* **164** (2): 501-527, 1874; *Chemical News* **29**: 1-3, 1874; **31**: 1-3, 11-13, 23-24, 33-35, 43-45, 53-54, 1875.

CROOKES, William. On attraction and repulsion accompanying radiation. *London, Edinburgh and Dublin Philosophical Magazine* **48**: 81-95, 1874; *Proceedings of the Physical Society of London* **1**: 35-51, 1876.

CROOKES, William. On repulsion resulting from Radiation. *Philosophical Transactions of the Royal Society of London*, **165**: 519-547, 1875; **166**: 325-354; 355-376, 1876; **170**: 87-134, 1879.

CURIE, Pierre; LABORDE, Albert. Sur la chaleur dégagée spontanément par les sels de radium. *Comptes Rendus des Séances de l'Académie des Sciences de Paris* **136**: 673-5, 1903.

DORN, Ernst. Versuch über die zeitliche Gewichtsänderung von Radium. *Physikalische Zeitschriften* **4**: 530-531, 1903.

DROSSBACH, Gerard Paul. On periodical fluctuations of the intensity of the Earth's gravity, and their influence on determinations of atomic weights. *Chemical News* **72**: 68, 1895.

EDDINGTON, Arthur Stanley. Majorana's theory of gravitation. *Astrophysical Journal* **56**: 71-72, 1922.

EÖTVÖS, Roland von; PEKÁR, Desiderius; FEKETE, Eugen. Beiträge zum Gesetze der Proportionalität von Trägheit und Gravität. *Annalen der Physik* [4] **68**: 11-66, 1922.

ERISMANN, Theodor. Zur Frage nach der Abhängigkeit der Gravitationskraft von Zwischenmedium. *Vierteljahresschrift der Naturforschende Gesellschaft zu Zürich* **53**: 157-185, 1908.

FÖPPL, August. Ein Versuch über die allgemeine Massenanziehung. *Physikalische Zeitschriften* **6**: 113-114, 1905.

FORCH, Carl. Bewirken radioaktive Substanzen eine Absorption von Gravitationsenergie? *Physikalische Zeitschriften* **4**: 318-319; 443-445, 1903.

FRESNEL, Augustin. Note sur la répulsion que les corps échouffés exercent les uns sur les autres à des distances sensibles. *Annales de Chimie et Physique* **29**: 57-61, 107-8, 1825. Reproduced in Vol. II, p. 667-672, in: *Oeuvres complètes d'Augustin Fresnel*. Ed. H. Senarmont, É Verdet & L. Fresnel. Paris: Imprimérie Impériale, 1868.

FRIEDLÄNDER, Benedict; FRIEDLÄNDER, Immanuel. *Absolute oder relative Bewegung?* Berlin, 1896.

GASPARI, E. Les études récents sur le pendule. *Revue Générale des Sciences Pures et Appliquées*, **6**: 401-407, 1895.

GEIGEL, Robert. Über Absorption von Gravitationsenergie durch radioaktive Substanz. *Annalen der Physik* [4] **10**: 429-435, 1903.

GOUDEY, Russel. Sur une variation périodique anuelle de la marche des pendules. *Comptes Rendus Hebdomadaires des Séances de l'Académie des Sciences de Paris* **175**: 748-750, 1922.

HARTSOUGH, Ralph C. Effect of lunar gravity upon a quartz thread balance. *Physical Review* [2] **19**: 282-3, 1922.

HEYDWEILLER, Adolf. Über Gewichtsänderungen bei chemischer und physikalischer Umsetzung. *Physikalische Zeitschriften* **1**: 527-9, 1900.

HEYDWEILLER, Adolf. Über Gewichtsänderungen bei chemischer und physikalischer Umsetzung. *Annalen der Physik* [4] **5**: 394-420, 1901.

HEYDWEILLER, Adolf. Bemerkungen zu den Gewichtsänderungen bei chemischer und physikalische Umsetzung. *Physikalische Zeitschriften* **3**: 425-426, 1902.

HEYDWEILLER, Adolf. Zeitliche Gewichtsänderungen radioaktiver Substanz. *Physikalische Zeitschriften* **4**: 81-82, 1902.

HEYL, Paul R. Gravitational anisotropy in crystals. *Scientific Papers of the Bureau of Standards* **19** (482): 307-24, 1924.

HICKS, William Metchinson. On some irregularities in the value of the mean density of the Earth, as determined by Baily. *Proceedings of the Cambridge Philosophical Society* **5**: 156-161, 1883-6.

JOLY, John. On the conservation of mass. *Transactions of the Royal Dublin Society* **8** (2): 23-52, plate V, 1903.

KAUFMANN, Walter. Bemerkungen zu der Arbeit des Hrn. R. Geigel: "Über die Absorption von Gravitationsenergie durch radioaktive Substanz". *Annalen der Physik* [4] **10**: 894-896, 1903.

KENNARD, Earle H. Electrical action and the gravitational constant. *Science* [2] **43**: 928-929, 1916.

KOCH, K. R. Über Beobachtungen, welche eine zeigliche Änderung der Grösse der Schwerkraft wahrschenlich machen. *Annalen der Physik* [4] **15**: 146-156, 1904.

KREICHGAUER, Damian. Einige Versuche über die Schwere. *Verhandlungen der physikalischen Gesellschaft zu Berlin* **10**: 13-16, 1891.

LAAGER, Fritz. *Versuch mit der Drehwage die Abhängigkeit der Gravitation von Zwischermedium nachzuweisen.* Bern: Haller, 1904.

LANDOLT, Hans. Untersuchen über etwaige Änderungen des Gesamtgewichtes chemisch sich umsetzender Körper. *Zeitschrift für physikalische Chemie* **12**: 1-34, 1893.

LANDOLT, Hans. Untersuchungen über die fraglichen Änderungen des Gesamtgewichtes chemisch sich umsetzender Körper. Zweite Mitteilung. *Sitzungsberichte der königl preussische Akademie der Wissenschaften zu Berlin* **14**: 266-298, 1906

LANDOLT, Hans. Untersuchungen über die fraglichen Änderungen des Gesamtgewichtes chemisch sich umsetzender Körper. Dritte Mitteilung. *Sitzungsberichte der königl Preussische Akademie der Wissenschaften zu Berlin* **16**: 354-387, 1908;

LLOYD, Morton G. Does magnetization alter mass? *Terrestrial Magnetism and Atmosferic Electricity* **14**: 67-71, 1909.

LO SURDO, Antonino. Sulle pretese variazioni di peso in alcune reazioni chimiche. *Nuovo Cimento* [5] **8**: 45-67, 1904

LODGE, Oliver Joseph. The discovery of radioactivity and its influence on the course of physical science. *Journal of the Chemical Society* **101**: 2005-2031, 1912.

MCGUIRE, James E; TAMNY, Martin. *Certain philosophical questions: Newton's Trinity notebook.* Cambridge: Cambridge University Press, 1983.

MACKENZIE, Arthur Stanley. On the attractions of crystalline and isotropic masses at small distances. *Physical Review* **2**: 321-343, 1895.

MAJORANA, Quirino. Sulla gravitazione. *Atti della Reale Accademia dei Lincei. Rendiconti. Classe di Scienze Fisiche, Matematiche e Naturali* [5] **28** (2° Semestre): 165-174, 221-223, 313-317, 416-421, 480-489, 1919; **29** (1° Semestre): 23-32, 90-99, 163-169, 235-240, 1920 (b)

MAJORANA, Quirino. Sur la gravitation. *Comptes Rendus des Séances de l'Académie des Sciences de Paris* **169**: 646-649, 1919 (a).

MAJORANA, Quirino. Expériences sur la gravitation. *Comptes Rendus des Séances de l'Académie des Sciences de Paris* **169**: 719-721, 1919 (b).

MAJORANA, Quirino. On gravitation. Theoretical and experimental researches. *London, Edinburgh and Dublin Philosophical Magazine* [6] **39**: 488-504, 1920.

MAJORANA, Quirino. Sur l'absorption de la gravitation. *Comptes Rendus des Séances de l'Académie des Sciences de Paris* **173**: 478-479, 1921.

MAJORANA, Quirino. Quelquer recherches sur l'absorption de la gravitation par la matière. *Journal de Physique et le Radium* [7] **1**: 314-323, 1930.

MANLEY, John Job. On the apparent change in weight during chemical reaction. *Philosophical Transactions of the Royal Society of London* **A 212**: 227-60, 1913.

MARTINS, Roberto de Andrade. Pesquisas sobre a absorção da gravidade. Pp. 198-213, in: *Anais do I Seminário Brasileiro de História da Ciência e Tecnologia.* Rio de Janeiro: Museu de Astronomia e Ciências Afins, 1986.

MARTINS, Roberto de Andrade. Os experimentos de Landolt sobre a conservação da massa. *Química Nova* **16** (5): 481-90, 1993.

MARTINS, Roberto de Andrade. Estudos experimentais sobre gravitação no início do século XX. Pp. 393-401, in: *VI Seminário Nacional de História da Ciência e da Tecnologia. Anais.* Rio de Janeiro: Sociedade Brasileira de História da Ciência, 1997.

MARTINS, Roberto de Andrade. The search for gravitational absorption in the early 20th century. Pp. 3-44, in: GOEMMER, H., RENN, J., & RITTER, J. (eds.). *The expanding worlds of general relativity (Einstein Studies, vol. 7).* Boston: Birkhäuser, 1999.

MARTINS, Roberto de Andrade. Majorana's experiments on gravitational absorption. Pp. 219-238 in: EDWARDS, Matthews R. (ed.). *Pushing Gravity: New Perspectives on Le Sage's Theory of Gravitation.* Montreal: Apeiron, 2002 (a).

MARTINS, Roberto de Andrade. Gravitational absorption according to the hypotheses of Le Sage and Majorana. Pp. 239-258, in: EDWARDS, Matthews R. (ed.). *Pushing Gravity: New Perspectives on Le Sage's Theory of Gravitation.* Montreal: Apeiron, 2002(b).

MARTINS, Roberto de Andrade. Opyty Majorany po poglosceniju gravitacii. Pp. 76-99, in: IVANOVA, M. A. &

SAVROVA, P. A. (eds.). *Poiski mekhanizma gravitacii*. Nizhnij Novgorod: Izd. Ju. A. Nikolaev, 2004.[9]

MARTINS, Roberto de Andrade. Searching for the ether: Leopold Courvoisier's attempts to measure the absolute velocity of the solar system. *Dio: The International Journal of Scientific History*, **17**: 3-33, 2011.

MARTINS, Roberto de Andrade. Émile Meyerson and mass conservation in chemical reactions: *a priori* expectations versus experimental tests. *Foundations of Chemistry*, **21** (1): 109-124, 2019.

NEWCOMB, Simon. Fluctuations in the Moon's mean motion. *Monthly Notices of the Royal Astronomical Society* **69**: 164-169, 1909.

NIPHER, Francis Eugene. Gravitation and electrical action. *Science* [2] **43**: 472-473, 1916; *Transactions of the Academy of Sciences of Saint Louis* **23**: 163-75, 1916.

NIPHER, Francis Eugene. Gravitational repulsion. *Science* [2] **46**: 293-294, 1917.

PLANCK, Max. Zur Dynamik bewegter Systeme. *Sitzungsberichte den preussische Akademie der Wissenschaften zu Berlin* **29**: 542-70, 1907. Reproduced in: *Annalen der Physik* [4] **26**: 1-40, 1908.

POTTER, Harold Herbert. Note on the gravitational acceleration of bismuth. *Physical Review* **19**: 187-8, 1922.

POTTER, Harold Herbert. Some experiments on the proportionality of mass and weight. *Proceedings of the Royal Society of London* **A 104**: 588-610, 1923.

POTTER, Harold Herbert. On the proportionality of mass and weight. *Proceedings of the Royal Society of London* **A 113**: 731-732, 1927.

POYNTING, John Henry. Recent studies in gravitation. *Nature* **62**: 403-408, 1900.

POYNTING, John Henry; GRAY, Peter L. An experiment in search of a directive action of one quartz crystal on another.

---

[9] Russian translation of Martins 2002a.

Philosophical Transactions of the Royal Society of London **A 192**: 245-256, 1899.

POYNTING, John Henry; PHILLIPS, Percy. An experiment with the balance to find if change of temperature has any effect upon weight. *Proceedings of the Royal Society of London* **A 76**: 445-457, 1905.

RAYLEIGH, Lord. Does chemical transformation influence weight? *Nature* **64**: 181, 1901; **66**: 58-59, 1902.

RICHARDSON, Owen Willans. Note on gravitation. *London, Edinburgh and Dublin Philosophical Magazine* [6] **53**: 138-143, 1922.

RUMFORD, Count (Benjamin Thompson). An inquiry concerning the weight ascribed to heat. *Philosophical Transactions of the Royal Society of London* **89**: 179-194, 1799. Reproduced in: *Journal of Natural Philosophy, Chemistry and the Arts* **3**: 381-390, 1799/1800; *Bibliothèque Britannique* **13**: 217-238, 1800.

RUSSELL, Henry Norris. On Majorana's theory of gravitation. *Astrophysical Journal* **54**: 334-346, 1921.

SAGNAC, Georges M. M. Une relation possible entre la radioactivité et la gravitation. *Journal de Physique Théorique et Appliquée* [4] **5**: 455-462, 1906.

SANFORD, Fernando; RAY, Lilian E. On a possible change of weight in chemical reactions. *Physical Review* **5**: 247-53, 1897; **7**: 236-8, 1898.

SEELIGER, Hugo von. Über die Anwendung der Naturgesetze auf das Universum. *Sitzungsberichte der Königlich Bayerischen Akademie der Wisenschaften zu München. Mathematisch-physikalische Klasse* (4. Abhandlung): 1-25, 1909.

SHAW, Philip E. The Newtonian constant of gravitation as affected by temperature. *Philosophical Transactions of the Royal Society of London* **A 216**: 349-92, 1916.

SHAW, Philip E.; DAVY, N. The effect of temperature on gravitative attraction. *Physical Review* **21**: 680-691, 1923.

SHAW, Philip E.; HAYES, C. A special test of the temperature effect of gravitation. *Proceedings of the Physical Society*, **29**: 163-169, 1917.

SIMONS, L. Appendix to RICHARDSON, O. W. Note on gravitation. *London, Edinburgh and Dublin Philosophical Magazine*, [6] **43**: 143-145, 1922.

SITTER, Willen de. Absorption of gravitation. *The Observatory*, **35**: 387-393, 1912.

SITTER, Willen de. On absorption of gravitation and the Moon's longitude. *Proceedings of the Royal Academy of Amsterdam*, **15**: 808-824, 824-830, 1913.

SOUTHERNS, Leonard. Experimental investigation as to dependence of gravity on temperature. *Proceedings of the Royal Society of London*, **A 78**: 392-403, 1907.

SOUTHERNS, Leonard. Experimental researches upon the dependence of the weight of a body on its state of electrification. *Proceedings of the Cambridge Philosophical Society*, **15**: 352-72, 1909.

SOUTHERNS, Leonard. A determination of the ratio of mass to weight for a radioactive substance. *Proceedings of the Royal Society of London*, **A 84**: 325-44, 1911.

THOMSON, Joseph John. Address of the President of the British Association for the Advancement of Science. *Science*, **30** (765): 257-279, 1909.

TOMASCHEK, Rudolf Karl Anton; SCHAFFERNICHT, Walter. Untersuchungen über die zeitlichen Änderungen der Schwerkraft. *Annalen der Physik*, [5] **15**: 787-824, 1932.

TOMASCHEK, Rudolf Karl Anton; SCHAFFERNICHT, Walter. Tidal oscillations of gravity. *Nature*, **130**: 165-166, 1932.

WALKER, Miles; STAINER, W. Witcomb. An inquiry into the possible existence of mutual induction between masses. *London, Edinburgh and Dublin Philosophical Magazine*, [6] **32**: 592-600, 1916.

WILSON, Harold Albert. Note on the ratio of mass to weight for bismuth and aluminium. *Physical Review*, **20**: 75-7, 1922.

WOODWARD, James F. *The search for a mechanism*. PhD Dissertation in History. University of Denver, 1972. (*Dissertation Abstracts International* **33**: 3515-A, 1973, UMI 72-33077)

ZEEMAN, Peter. Some experiments on gravitation. The ratio of mass to weight for crystals and radioactive substances. *Proceedings of the Royal Academy of Amsterdam*, **20**: 542-553, 1918.

ZENGHELIS, Constantin. Zur Frage der Erhaltung des Gewichtes. *Zeitschrift für physikalische Chemie* **65**: 341-358, 1909.

THE SEARCH FOR AN INFLUENCE OF TEMPERATURE ON GRAVITATION

Roberto de Andrade Martins

Abstract: There were several careful experimental attempts to find a temperature effect upon gravitational force, from Lavoisier and Rumford (towards the end of the 18th century) to Shaw (20th century). Some positive effects were sometimes reported (Hick, Shaw) but further investigation showed that no influence seemed to exist. For temperatures up to 200 °C, the searched effect was established to be less than $10^{-9}/°C$ for passive gravitational mass (weight) and less than $10^{-6}/°C$ for active gravitational mass (attraction). This article describes the history of this search and discusses methodological issues raised by it.
Keywords: gravitation; non-Newtonian effects; experiments on gravitation; temperature and gravitation

1. INTRODUCTION

When one thinks about experimental or observational gravitational research made at the beginning of the 20th century, one usually recalls general relativity and its three "classical tests": perihelion precession, red-shift and light deflection. Nevertheless, at the turn of the century, there was an intensive research on gravitation (both theoretical and experimental) providing exciting ideas and unexpected empirical results. Only after the success of general relativity, in the 1920's, those independent lines of research subsided to the background,

MARTINS, Roberto de Andrade. *Studies in History and Philosophy of Science I*. Extrema: Quamcumque Editum, 2021.

rapidly disappearing (but for sporadic articles) towards 1930. The general outlook of those researches may be learned from several review papers of that time (Drude, 1897; Zenneck, 1901; Poincaré, 1953; Oppenheim, 1920; Poynting, 1900).

The aim of this paper is to provide an analysis of one of those forgotten lines of research: the attempt to find a connection between gravitation and temperature. The article will focus upon experimental work, leaving aside most of the bulky speculative material of the time.[1]

Nowadays, we have learned from the theory of relativity that heating a body will increase its mass, since any energy change ΔE of a system will produce a mass change $\Delta m = \Delta E/c^2$. As a matter of fact, a theoretical prediction of the relationship between heat and inertial mass[2] had been made by Friedrich Hasenöhrl (1874-1915), before the researches of Albert Einstein (1879-1955) on this subject (Hasenöhrl, 1904-1905). The influence of temperature upon weight was discussed for the first time by Max Planck (1858-1947) (Planck, 1907) and it stimulated the first Einsteinian attempt to study gravitation (Einstein, 1907). According to the theory of relativity, an increase of 50 °C of the temperature of 1 kg of water will

---

[1] This work was written during the years 1995-1996, while the author was a visiting scholar of the Department of History and Philosophy of Science, University of Cambridge; and a visiting fellow of Wolfson College. It is now published for the first time. The content of the paper has not been updated or complemented, only slight changes were made.

[2] It is important to remark that there are several distinct mass concepts. Inertial mass is the quantity that appears in the dynamical laws of mechanics $(p=m.v)$. Gravitational mass is the quantity related to gravitational forces. Two kinds are now distinguished: active gravitational mass is the source of the gravitational field; passive gravitational mass is the quantity in the attracted body that reacts to the external gravitational field. In principle, one of those masses may change without any change of the others. See Reichenbächer, 1923, Bondi, 1957.

produce a mass increase of $2,3 \times 10^{-12}$ kg – a difference impossible to detect with any available instrument, even now. But the attempts to measure gravitational effects due to heat or temperature did not spring from the theory of relativity: they came either from different theoretical sources or from mere guessing: "what will happen if...". They were part of a broader exploratory approach to experimental gravitation that was especially strong in the late nineteenth and early twentieth centuries.[3]

Although those attempts led to no discovery, they have some relevance for several reasons: (1) they did show that there was no detectable influence of heat or temperature upon gravitation (and negative results are important scientific data); (2) they seemed to yield positive results for some time – a result that stirred the scientific community; and (3) they help us to learn a forgotten chapter of gravitational research.

2. EARLY IDEAS AND EXPERIMENTS TO THE END OF THE 18TH CENTURY

Within the context of Aristotelian physics, heat and cold are respectively linked to lightness and weight. Fire and hot air tend to ascend and are therefore light. This was, perhaps, the oldest proposed correlation between heat and gravitation (weight). Even after the decay of Aristotelian physics and the acknowledgment that hot air was not absolutely light but got its tendency to go up from the surrounding air, this deeply seated and intuitive idea of a negative weight related to heat did not disappear altogether: it reappears, for instance, inside flogiston theories, in the 18th century (see Partington & McKie, 1937-1939).

Another line of thought linking heat to repulsion may be found in Newtonian physics. Isaac Newton (1643-1727) did not

---

[3] See my previous paper in this volume: Roberto de Andrade Martins, Experimental studies on mass and gravitation in the early twentieth century: the search for non-Newtonian effects.

accept the Baconian conception of heat as a hidden vibrating motion of the smaller parts of matter. Instead, he speculated about the idea of heat as a repulsive force between atoms. In the second book of Newton's *Principia*, he was able to explain Boyle's pressure-volume relation using a model of static repealing atoms. In this model, these forces were not gravitation: they were short ranged and varied inversely as the distance. Nevertheless, here we find another deeply rooted and intuitive idea (that sometimes spontaneously arises in the minds of physics students): heat as a repulsive force. This conception is able to explain the dilation of heated bodies and might suggest that highly heated bodies might repeal the Earth and would become "light".

A third important conception is that of heat as a substance[4]. This idea is suggested by the possibility of transferring heat from a body to another and by calorimeter experiments where there seems to be a quantitative conservation of heat – and, remember, the main mark of (Aristotelian) substances, as material causes, was their conservation. Antoine-Laurent Lavoisier (1743-1794) and Pierre-Simon de Laplace (1749-1827) thought that heat (caloric) was a substance, although they did not exclude the possibility of its being just a kind of internal motion (Lavoisier & Laplace, 1780). But the conception of a substance leads to the property of weight – and, indeed, Lavoisier asked himself, about 1770, whether a body would become heavier when hot. In one of his laboratory notebooks (Berthelot, 1902) he referred to Georges-Louis Leclerc de Buffon (1707-1788), who "seems to have proved by experiments deemed by him conclusive, that fire matter weights, and that an incandescent body has from 1/350 to 1/600 of it [fire] in it mass". Nevertheless, Lavoisier stated that Herman Boerhaave (1668-1738) had already shown that an iron

------

[4] For a review of heat conceptions towards the end of the 18th century, see Bentham, 1937.

rod does not change its weight when becoming incandescent – and he accepts Boerhaave's result.

At the end of the 18th century, the problem of weight (or lightness) of heat was much discussed, mainly in the context of flogiston theory. The experimental problem was delicate: heated (calcinated) metal increases in weight, even after cooling. This effect was known in the 17th century and was explained by Robert Boyle as due to the weight of heat. It seems that this was a popular subject of inquiry at that time, as we find a description of experiments on [5]the increase of the weight of metals by fire in the Royal Society. We now ascribe this effect to absorption of oxygen and oxide formation, but the role of oxygen was completely understood only after Lavoisier's researches.

Besides that, from a purely physical point of view, a heated body will seem lighter than the same body cold, for three main reasons: i) ascending convection currents of the surrounding colder air, that produce an upwards force upon the body; ii) volume increment that produces a greater upwards aerostatic thrust upon the body; and iii) changes of the moisture attached to the surface of the bodies. For the same reason, a colder body will show heavier than the same body heated. None of those effects will be observed if weight measurements are made in a vacuum – but vacuum weighing is very difficult.

The effect of air currents was studied in the early 17th century by the *Accademia del Cimento* (Tozzetti, 1780, vol. 2.2, pp. 618-619). Some experiments had shown that a hot piece of metal seemed lighter; but it was then observed that placing an incandescent piece of iron *near* the balance would also produce an apparent weight decrease of anything placed at the balance, and it was then easy to find the cause of this effect.

2.1 Lavoisier's ice experiment

Leaving aside several earlier discussions, let us review some experimental attempts to study the problem made at the end of

---

[5] See Sprat, *The history of the Royal Society of London*, pp. 228-9: "Experiments on the weight of bodies increased in the fire".

the 18th century. The subject is discussed by Lavoisier (with the collaboration of Laplace) in a paper presented to the French Academy of Sciences in 1783 (Lavoisier, 1783). This work discusses the change of weight of phosphor and sulfur when they are burned. The weight increase is ascribed by Lavoisier to part of the air (oxygen) that combines with phosphor or sulfur; but another explanation is discussed: could not this change be ascribed to the heat that escapes from those substances when they are burned? In this case, the weight of heat should be negative, of course. In order to test this idea, the following experiment is presented: 8 grains of phosphor are burned in a sealed flask; the weight of the closed flask is the same both before and after the combustion (after it is allowed to become cold). The precision of the balance is stated to be 1/4 of grain. Supposing that the whole of the phosphor did really burn, this test shows that the weight change was smaller than 1/32 (or 3%) of the weight. For our standards, this is a poor test, but it was highly relevant, in the context of the experiment, since the changes of weight of sulphur and phosphor, when burned, could be up to 100%.

After this chemical test, they try another approach. In order to test whether heat has any weight, they look for a weight change of a flask with water when the water is frozen. From previous measurements, they had already found that the burning of 92 grains of phosphor will be able to melt 1 pound of ice. Therefore this new experiment (much safer than the burning of a large amount of phosphor in a closed flask) can be readily used to discuss the weight changes of phosphor.

The ice experiment is produced in the following way: a glass flask is filled with 1 pound of water close to its freezing point. The flask is hermetically closed and its weight is found. The flask is then put into a snow and salt bath until the water freezes. The flask is now weighed again. The ice is then melted and the process is repeated several times. No change of weight is observed. A warning is presented that the environment must be at a temperature close to the freezing point, in order to avoid condensation of moisture. As the water is always at similar temperatures, no problem arises from convection currents or changes of volume of the flask.

The sensibility of the test is stated to be of 0.1 grain. As one pound corresponds to 9,216 grains, the change of weight of water as it froze or melted was less than 1/92,000 or about 1,1 x 10^{-5}.

2.2 Fordyce's ice experiment

A similar experiment was made by George Fordyce (1736-1802) (Fordyce, 1785). It was not possible to detect whether he knew Lavoisier's experiment.

Fordyce took a glass vessel (its weight was 451 grains) and put inside the vessel 1,700 grains of water, leaving part of the vessel full of air. The vessel was hermetically sealed. Its initial temperature was 37° F. It was put in a refrigerating mixture at 12° F until it began to freeze. Fordyce took the vessel of the mixture, shook and carefully wiped it and measured its weight. The vessel was then put again in the refrigerating mixture, and Fordyce made repeated measurements of its weight as the freezing went on. The weight was found to increase steadily. When all the water had been turned to ice, the system had gained close to 1/5 of a grain. The temperature of the ice inside the vessel was now about the same as the temperature of the refrigerating mixture (12° F). In order to avoid any effect due to temperature differences, Fordyce took the vessel out of the mixture and waited until it began to melt (that is, until its temperature became 32° F). Repeated measurements showed that the weight decreased. When the temperature reached 32° F, the total weight gain, since the beginning of solidification, was about 1/16 of a grain. He waited until all the ice had thawed, and found that the weight returned to its initial value (at the beginning of solidification) but for a very slight increase of about 1/1600 of a grain (one division of the scale of his balance). The reported change was about 1/27,000 or 3,5 x 10^{-5} of the weight of the water.

Fordyce says that the experiment was repeated several times, always with similar results. In the discussion of the effect, he stated:

> The acquisition of weight found on water's being converted into ice, may arise from an increase of the alteration of gravitation of the matter of the water; or from some substance imbibed through the glass, which is necessary to render the water solid.

Fordyce recalls that heat decreases some small-scale forces (cohesion, chemical forces) and suggests an interesting experiment to test whether there was only a force change or a matter transport. In modern terminology it may be described thus: Suppose we have two equal hollow pendulums, one full of ice and the other with an equal mass of water. The period T of a pendulum may be computed as

$$T = 2\pi\sqrt{\frac{L.m}{w}}$$

where L is its length, m the (inertial) mass of the swinging body and w its weight. In common cases, this formula reduces to

$$T = 2\pi\sqrt{\frac{L}{g}}$$

where g is the gravitational acceleration. But this formula is equivalent to the previous one only if $w = mg$, and this relation could be violated in some cases (if heat decreases the gravitational force without changing the inertial mass). So, if heat decreases the weight without transference of matter, the period of the water pendulum will be greater than the period of the ice pendulum. But if there is any transference of matter in the process, weight changes will be accompanied by proportional inertia changes and the period of the pendulums will be equal. Fordyce even considers the possibility of "absolutely light" matter, that would have inertia but would repel common matter, instead of attracting it. In that case, the change of the period would be twice as large, because a weight increase would be associated to an inertial decrease and both changes would add instead of compensating each other. In modern terminology, Fordyce is proposing a test of the principle of equivalence (or of proportionality of inertial and gravitational mass) comparing water an ice.

Although Fordyce makes some slips in his argument, the essential idea (as exposed here) is correct and very interesting. Nevertheless he does not make the test – and it would not be easy to detect period changes of 10^{-5} at that time.

2.3 Rumford's repetition of Fordyce's experiment

Motivated by Fordyce's previous experiments, Benjamin Thompson, Count Rumford (1753-1814) began in 1787 to study the question (Thompson, 1799). His method was carefully devised in order to avoid spurious results.

He took two very similar flasks that he named A and B. He put into A a weighted amount of water (4107.86 grains Troy), leaving about half the bottle empty; in the other flask (B) he put an equivalent weight of "weak spirit of wine" (alcohol). His clever idea was that he could produce the freezing of the water at a temperature such that the alcohol would not freeze; the heat leaving the water when it freezes would be much greater than the heat leaving the alcohol, but all other effects would be highly similar. Comparing the two flasks, it seemed possible to measure any weight effect due to heat.

Both flasks were hermetically sealed, washed, cleaned and dried. They were suspended at the two ends of the arms of a fine balance, inside a room heated to 61°F. After he thought the bottles and their contents had reached this temperature, he equilibrated the balance by using a small silver wire. He waited for 12 hours and observed that there was no alteration of the equilibrium. Then, he removed the whole apparatus to an unheated room (it was winter) where the temperature was 29° F. he left the balance and bottles there for 48 hours and, after this time, he observed that the balance now inclined towards the flask containing water (*A*). This seemed to show that the weight of the water had increased (or the weight of the spirit of wine had decreased). As the greater heat effect should happen with the water (because of its change of state), it seemed that the water had increased its weight.

The weight difference between the flasks was now 0.134 grains (that is, 1/35,904 of the total weight) – an effect of the same order of magnitude as that measured by Fordyce. The water had completely frozen, and the spirit of wine was still liquid. Therefore, a much larger amount of heat had gone out of the water than from the other flask and the observed effect should be more likely ascribed to the water: by losing heat, it became heavier!

Notice that the experiment does indeed avoid spurious effects due to convective currents, water attached to the surface of the flasks and volume changes. The effect seemed real, but Rumford was very surprised by the result and did not publish it. He carefully studied the balance itself, and found no defect on it. He repeated the experiment and observed that the change of weight was reversible; but he noticed that the amount of weight change was not always the same. Repeating the experiment with water and mercury, no effect was observed.

At last, Rumford found out that the observed effect was due to two main reasons: small initial temperature differences between the water and the spirit of wine; and problems associated to moisture at the surface of the flasks. By a more careful repetition of the experiment, the effect was eliminated – there was no weight difference, at a stated sensitivity of 1/1,000,000 or 10^{-6}. Rumford boldly concluded that "all attempts to discover any effect of heat upon the apparent weight of bodies will be fruitless".

As Rumford himself states in his article, the experiment was a relevant test about the hypothesis of substantiality of heat. As he was already convinced that heat was a kind of motion, he was led to repeat the experiment until he found out its defects. Had he believed heat was a substance, he would probably have published his first results, since they were seemingly free of spurious effects.

3. RADIATION REPULSION: FRESNEL AND CROOKES

Several attempts were made during the 19th century to detect a repulsive effect of heat or radiation. Augustin-Jean Fresnel (1788-1827) attached light bodies to the ends of a torsion balance and concentrated the light of the sun upon one of them, by means of a lens (Fresnel, 1825). He noticed no effect. But, when he repeated the experiment inside an evacuated container, he noticed an apparent strong repulsion between two test bodies heated by the sun's light. The air pressure was about 1 or 2 mm of mercury. In order to test whether the effect was due to the remaining air, Fresnel increased the pressure to 20 mm of mercury and the repulsion became much weaker. He concluded that the observed repulsion was not produced by the air. Being unable to explain it, Fresnel did not continue his research[6].

Some strong effects of air currents produced by heating were sometimes observed and explained by Claude Servais Mathias Pouillet (1791-1868) (Pouillet, 1849). But Fresnel's effect was of a different kind.

This effect was elucidated half a century later by William Crookes (1832-1919). In order to make some delicate weighing to measure the atomic weight of thallium (Crookes, 1873), Crookes built a balance that could be operated in a vacuum. He expected it to be free from every systematic error produced by heat, as there would be no air convection or atmospheric thrust. Nevertheless, the balance seemed to show that bodies were lighter when hot than when they were cold. Crookes conjectured

---

[6] It seems that a similar experiment was made at the *Accademia del Cimento*, as one may infer from a drawing that, unfortunately, was not accompanied by any written description or explanation. The drawing will be found in Tozzetti, 1780, vol. 3, fig. 304.

that this effect could be due to a repulsive force produced by heat. In order to test this hypothesis, he used a torsion balance inside an evacuated container, as Fresnel had done before him (Crookes, 1874).

Crookes observed an apparent repulsion between hot bodies using this king of apparatus. Even placing a hot body outside the glass container, the same effect was observed. A cold body produced an apparent attraction. At the beginning, Crookes believed that this was a direct effect produced by the hot body itself; but he noticed that the effect varied as he changed the pressure of the residual gas in the container. At last, his continued search led to the discovery that the effect was due to the remaining air itself, (Crookes, 1875-1876; Schuster, 1876; Reynolds, 1876) and it was named "radiometer effect". Crookes' radiometer is now well known and it is not necessary to describe it here. Notice that this effect is completely different and much larger than the direct effect of radiation pressure that was measured some years latter by Pyotr Nikolayevich Lebedev, and by Ernest Fox Nichols and Gordon Ferrie Hull (see Schagrin, 1974).

During the whole of the 19th century, several other naïve attempts were made to detect some kind of temperature attraction or repulsion. Crookes himself describes several of these (Crookes, 1875). Besides the previously described ideas that could lead somebody to conjecture about the existence of such effects, there were new reasons, now. The 19th century was the period when ether theories grew and expanded to all fields. After their success in optics (the revival of the wave theory of light), electricity and magnetism (Faraday's and Maxwell's theories, among others), they also conquered gravitational theoretical thought.

Now, if gravitation is not a direct interaction at a distance but something transmitted by a medium, one might speculate that the interaction between the bodies and the medium or the properties of the medium itself could change as a function of temperature. This was strongly suggested by the so-called "kinetic theories of gravitation", where the gravitational ether was analogous to the kinetic gas model (see specially Taylor, 1876). This theoretical background is a possible explanation of the increased interest in experimental gravitational research, specially towards the end of the 19th century and beginning of the 20th century.

It is interesting to remark that there was also some indirect evidence of an influence of temperature on gravitation. In the *Principia*, Newton developed a mechanical model for reflection and refraction, and in the *Optics* suggested a possible relationship between the forces that bend light rays and gravitational forces. As the index or refraction of glass changes with temperature, John Herapath suggested that heat changes gravitation. "In my own mind I have no doubt of the fact, but we are deficient of direct experiments to prove it." (Herapath, 1847, vol. 1, p. xv).

4. TEMPERATURE ANOMALIES IN MEASUREMENTS OF THE GRAVITATIONAL CONSTANT

The next hint about a possible relation between temperature and gravitation was provided by standard experiments designed to measure the so-called Newtonian constant of gravitation G. In the 19th century, such measurement was usually described as a measurement of the mean density d of the Earth – because one of these parameters can be calculated from the other, provided one knows the gravitational acceleration g and the radius of the Earth R:

$$g - GM/R^2 - G(4/3)\pi R^3 d/R^2 - (4/3)\pi dGR$$

4.1 Cornu and Baille

In 1873, Alfred Cornu (1841-1902) and Jean-Baptiste Alexandre Baille (1841-1918) noticed a difference of about 1% between their measurements for G made in winter and in summer:

summer: $G = 6{,}760 \times 10^{-8}$ Nm²kg⁻² $d = 5{,}56$ g.cm⁻³
winter: $G = 6{,}836 \times 10^{-8}$ Nm²kg⁻² $d = 5{,}40$ g.cm⁻³

The authors explained this difference as due to a small deflection of the balance beam and ascribed no importance to this result (Cornu & Baille, 1873).

4.2 Hicks

Two years latter, William Mitchinson Hicks (1850-1934) analyzed Francis Baily's 1841-2 measurements of the gravitational constant (Baily, 1843; Hicks, 1883-6) in order to search for a temperature anomaly. Baily's experiments were made at temperatures varying from 30° F in winter to 69° F in summer. Hicks carefully classed a large number of observations by temperature ranges. He found a regular correspondence between temperature and the measured value of the mean density of the Earth, as shown in Table 1.

Temperature: (°F)	Number of observations:	Mean density of the Earth: (g.cm⁻³)
36	46	5.7296
40±2	128	5.7341
45±2	247	5.6823
50±2	302	5.6799
55±2	187	5.6594
60±2	333	5.6495
65±2	140	5.5935
68	96	5.5828

Table 1. Relation between temperature and computed density of the Earth, according to Hicks' analysis of Baily's data.

Hick's results seemed to disclose a regular increase of gravitational attraction with temperature rise. The observed difference between the values measured at the highest and lowest temperatures amounted to about 3% and could not be ascribed to statistical fluctuations.

Notice that Baily's data, used by Hicks, were obtained with a set of different materials; therefore, the temperature effect was not due to the peculiar behaviour of a particular substance.

Hicks began to search for possible sources of systematic error. He studied the effect of temperature on the density of the air and computed effects arising from this. He showed that taking these effects into account he could increase the coherence of Baily's results; but that affected only the fourth decimal place in his table. He studied other effects and concluded that none of them could explain the observed anomaly.

4.3 Analysis of the results

At this point it is convenient to analyze and compare some of the previously described experiments. Let us suppose that there were no systematic errors. What could one conclude? If Rumford did not observe any change of weight above 10^{-6}, couldn Hicks measure anything as great as 3%?

Rumford's experiment does not show that temperature has no influence on gravitation. It just showed that no sensible effect was noticed of *heat* differences between water and the other tested substance (spirit of wine) or mercury). But if weight depended only on *temperature*, according to some equation as:

$$F = F_0 . f(T)$$

Rumford's experiment would be unable to tell anything about $f(T)$. It would be necessary to compare a cold to a hot body, in order to notice any effect.

Notice also that in Rumford's experiments only the temperature of the attracted body changed. The attracting body (the Earth) was always at the same mean temperature. If there did exist a temperature dependence on attraction but if it were a function of both temperatures

$$F = F_0 . f(T_1, T_2)$$

it could happen that no effect would be observed when only one of the temperatures (and specially the temperature of the smaller body) is changed.

Now, in the case of Hick's work, both attracting bodies were at the same temperature. The effect was a very remarkable one (3%), while Rumford's result showed no (relative) change of weight greater than 10^{-6}. But the experiments are not comparable. Although they seem to clash, they are indeed compatible.

Finally, let us compare Hick's work to the measurements of Cornu and Baille. At first sight, they seem incompatible: if both effects are real, one shows a decrease and the other an increase of G as the temperature increases. However, Cornu and Baille's article does not provide detailed information and it is possible that their winter measurements were made in a heated room.

4.4 Geophysical studies

An apparent relationship between temperature and gravitation was fortuitously found in geophysical studies. After the development of accurate pendulum systems, gravity measurements were made as a routine at every place. In the 19th century, measurements were started inside deep pits, in order to detect the regular gravity decrease predicted by the Newtonian gravitational theory. Outside the Earth, gravity increases as we approach the ground; but inside the Earth, gravity decreases to zero as we approach the centre.

A strange result was however noticed: when three independent measurements made in Haton, Pribran and Freiberg were compared, it seemed that the gravity decreases were very different, in different pits, for the same depth (Sterneck, 1899). For each 100 m of depth increase, gravity diminished about 1.4×10^{-5} and 1.5×10^{-5} at Haton and Freiberg, but only 0.9×10^{-5} at Pribran. On the other side, temperature measurements were also made, and it was noticed that temperature varied much more inside the Freiberg pit (about 3.8° C for each 100 m) than at Pribran (about 1.4° C for each 100 m). As a matter of fact, there was a stronger correlation between gravity changes and temperature than between depth and gravity. It seemed that, inside the several pits, equal gravity corresponded to equal temperatures. Of course, the measurements were free from elementary systematic errors (beam dilation, etc.).

As this strange fact was noticed, the Vienna Academy of Sciences decided to study the problem, and Robert Daublebsky von Sterneck (1839-1910) was charged to measure gravity variations at four different pits (Sterneck, 1899). A slight correlation was found between temperature and gravity. For each 100 m depth increase there was a greater gravity decrease corresponding to greater temperature increases: a -4.3×10^{-5} / °C relative change in g. But, as the mean errors were about 3×10^{-5}, one could not ascribe much value to those results.

Later on, Sterneck's results were cited by Philip Shaw (Shaw, 1916) as providing an indirect evidence of an influence of temperature upon gravitational attraction. However, if the effects were not spurious, it would be necessary to suppose that slight temperature changes (a few degrees) could be able to produce enormous changes (about 5%) of the gravitational attraction of superficial layers of the Earth's crust. The previous results obtained with torsion balances was sensitive enough to rule out such effect.

5. BALANCE EXPERIMENTS: POYNTING AND SOUTHERNS

5.1 Poynting and Phillips

Hicks' work led, twenty years later, to two experimental searches for temperature influences on weight. The first one was done by John Henry Poynting (1852-1914) and Percy Phillips (c. 1880-c. 1825), in 1905. They dismissed Hicks' results as spurious, although without explaining them. After briefly describing the difficulties of repeating direct measurements of gravitational attraction (with torsion balances) at different temperatures, they choose another method: to search for a weight change when a body is heated or cooled (Poynting & Phillips, 1905).

The test body was attached to one arm of a balance; at the other side, a counterpoise was kept at a constant temperature. The difficult aspect of the experiment was to avoid spurious effects such as those above described: convection currents, radiometer pressure, and so on. The apparatus was carefully devised and deserves a short description (Fig. 1).

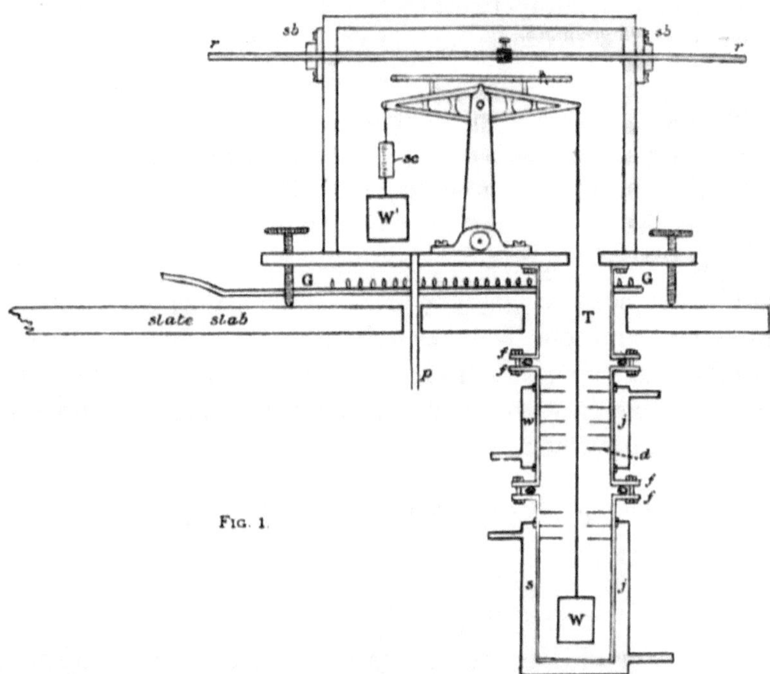

W, weight of which the temperature is to be raised, W' counterpoise.
T, tube in which it hangs, with a number of diaphragms with ½-inch holes.
sj, steam jacket, replaced by liquid air jacket.
ff, flanged joint with lead washer.
wj, water jacket.
p, pipe to the exhausting pump.
sc, scale read by a microscope not shown.
rr, rider rod passing through stuffing boxes, sb, enlarged in fig. 2.
gg, gas burners to heat the base plate before sealing up.

Fig. 1. The apparatus used by Poynting and Phillips (1905, p. 448).

At the left arm of the balance a solid gun metal cylinder W is hung by a short thin wire. SC is a scale that can be read by a microscope (not shown in the figure) in order to read the equilibrium position of the balance. From the right arm is hung the test body W, of the same weight (266 g) and material as W', hung to the balance by a long wire. The wire passes through several pierced screens designed to act as heat screens and convection holders.

The brass tube T where the test body W was enclosed was surrounded by two cooling (or heating) jackets. The upper one (wj) was a water jacket intended to keep the upper part of the tube at room temperature. The lower one (sj) was intended to produce temperature variations in W by passing either steam, or water, or liquid air.

The whole apparatus was air-tight and during the measurements the system was evacuated by the pipe P until the air pressure reached a fraction of a millimeter of mercury. This was done in order to avoid any spurious effects due to convection or air thrust.

The experiment consisted of observing eventual weight changes when the temperature of W was changed. When steam of liquid air were used to heat or cool down the tube T, large effects were observed, but they disappeared as soon as the temperature became constant. There remained a small weight decrease, when the test body was heated by steam (to $100°$ C): a mean change of -0.055 mg (of othe total 266 g). When it was cooled by liquid air (to $-186°$ C) the observed change was only 0.0016 mg. The authors suspected that the measured amount was due to some systematic error and tried to detect it using *hollow* test bodies: any superficial effect (such as residual air convection, radiation or radiometric pressure, etc.) should be the same with solid and hollow test bodies; but real weight changes should be different.

Changing W an W' by similar hollow cylinders (58 g) of the same volume, size and material as the first ones, the measurements were repeated. The observed weight changes were very similar to the preceding ones (-0.058 mg and 0.0007 mg, respectively). Therefore, if there was any real weight change it could amount to only 0.003 mg (when steam was used) or 0.001 mg (when liquid air was used). The authors conclude that, if there is any influence of temperature on weight, it must be less than one part in 10^9 per 1 ° C.

5.2 Southerns

The experiment made by Leonard Southerns (1878-1962), although published two years latter than Poynting's (Southerns, 1907), was an independent research, using an apparatus built by Hicks. There was one very important difference: instead of external heating of the test body, as used by Poynting and Phillips, Southerns used a well insulated weight that could be rapidly heated from inside, by an electric current. In this way, most of the spurious effects could be avoided, because the external surface of the body would become hot only after some time delay. The experiment was made in reduced pressure conditions.

No regular effects were observed, although some transient variations, during and shortly after heating, were observed. The temperature differences were smaller than in Poynting's experiments and the author concluded that if there is any influence of temperature upon weight, it is less than 1 part in 10^8 per 1° C. Although Poynting's work had produced a better result, Southerns' experiments are relevant as bringing independent confirmation of this null result.

6. THE DETECTION OF A POSITIVE EFFECT: SHAW

After all those attempts, it might seem that it was useless to make other tests. But Poynting's and Southerns' experiments could only change the temperature of the attracted test body – not of the attracting body (the Earth). Therefore, it was desirable to study a situation where the temperature of the attracting body could be controlled. This kind of test was done by Philip Egerton Shaw (1866-1949), and, as will be described, it led at first to the observation of a positive effect. His final results were stated thus: "The conclusion is that there is a temperature effect of gravitation. When one large mass attracts a small one, the gravitative force between them increases by about 1/500 as the temperature of the large mass rises from, say, 15 °C to 215 °C" (Shaw, 1916a).

Shaw's main article on this subject (Shaw, 1916a) was a massive 44 pages long paper published in the *Philosophical Transactions of the Royal Society* for 1916. The article was presented by Sir Charles Vernon Boys (1855-1944) – the man who had greatly improved the measurements of gravitational attraction, 20 years earlier (Boys, 1894). Both the names of Boys and of the Royal Society helped to give a great impact to this article.

Shaw's paper is very well written. It begins by discussing previous theoretical and experimental knowledge of the subject; describes his own careful method, the apparatus, observations and results; and presents a final discussion and conclusion.

Shaw provided an interesting review of previous evidence both for and against the temperature effect. Besides discussing the contributions of several of the authors referred to in the present paper, he analysed three indirect evidences:

i) There are two different methods of measuring G (or the mean density of the Earth): either by the use of the torsion balance, or by the measurement of the gravitational effect of large masses of the Earth's crust (either mountains, or making measurements in mine pits, for instance). Shaw pointed out that the temperature of mountain masses and superficial shells of the Earth's surface is above ordinary laboratory temperatures and that these "Earth" methods lead to a mean density of the Earth of about 5.4 g.cm^{-3}, while laboratory measurements lead to a value of 5.51 g.cm^{-3}. This difference could be interpreted as an *increase* of gravitational force with temperature.

ii) Shaw analysed Boys' experiments in the same way as Hicks had analyzed Baily's. He found a seemingly regular rise of G (or a decrease of the mean density d of the Earth) as the temperature increased. The change was about 1/1,000 for a rise in temperature of 1 °C.

iii) Measurements of G in mines, made by von Sterneck, had shown different results in mines with different temperature gradients. Shaw analysed the data and showed that they seem to point again to an increase of gravitation with temperature.

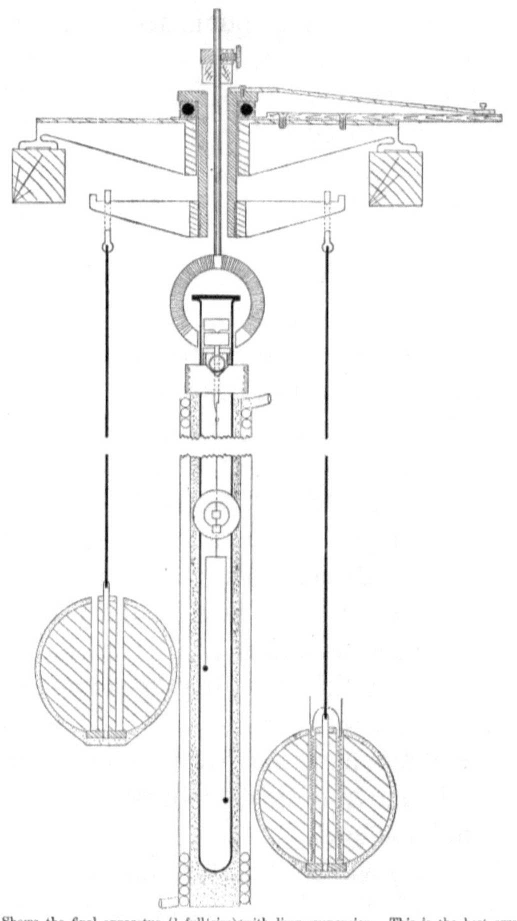

Fig. 2. The final form of the apparatus used by Shaw (1916a, p. 365).

Shaw decided to make an experiment measuring the force of attraction using a torsion balance (Fig. 2). The attracted masses were kept at the same temperature, inside an insulated case. Only the attracting bodies were heated and cooled. Their temperatures varied from room temperature (about 15 °C) to more than 200 °C. He carefully described several precautions and possible sources of error. He worked for eight years, improving the apparatus and making several tests. After a series of measurements, Shaw concluded that there is a regular increase of gravitational attraction of about $1.2 \times 10^{-5}/°C$, when the attracting body is heated.

7. REACTIONS TO SHAW'S RESULTS AND FURTHER EXPERIMENTS

A strong reaction against Shaw's results appeared very soon in *Nature*'s "Letters to the editor". From June 1916 to April 1917, when the discussion disappeared, 13 letters were published on this subject, by Joseph Larmor, Edwin Barton, Frederick Lindemann and Charles Burton, and George Todd, with replies by Shaw (Barton, 1916, 1917; Larmor, 1916a, 1916b, 1917; Lindemann & Burton, 1917; Lodge, 1917; Todd, 1917a, 1917b; Shaw, 1916b, 1917a, 1917b, 1917c). As it usually happens in those cases, the discussion was limited to theory: nobody criticized, made suggestions or repeated Shaw's experiments. The discussion exposed a general confusion between several mass concepts – specially those now called "passive" and "active" gravitational mass. At the end of the discussion, as at the very beginning, it was not clear whether Shaw's results were compatible or not with the basic laws of physics, or what change they would entail, if confirmed.

Meanwhile, Shaw, with the assistance of Cecil Hayes, repeated the experiment (Shaw & Hayes, 1917) in order to answer to a criticism presented at the meeting of the Royal Society, when the firs paper was read. It was suggested that convection currents around the large attracting body could change its position (towards the attracted bodies) and produce an apparent increase of gravitational force. A displacement of only 0.15 mm would be sufficient to account for the observed effect. In order to check this possible source of error, Shaw devised a method to measure the precise position of the attracting masses. He observed a displacement of the hot bodies – but it was an *outwards* displacement, of about 0.01 mm. This could not account for the observed effect. Indeed, this displacement showed that the effect was greater than that previously ascertained. As a result of this test, Shaw presented the result:

$$a = + (1.3 \pm 0.05) \times 10^{-5}/°C$$

for the temperature coefficient of gravitational attraction (a).

At the presentation of this second paper at the Physical Society, a few experimental doubts and suggestions were made by Charles Vernon Boys, Frederick Escreet Smith and Clifford C. Paterson, but they were easily replied by Shaw (Shaw & Hayes, 1917, p. 170). Nevertheless, Shaw perceived that a possible source of error was the displacement of either the large or the small masses – and decided to improve his apparatus.

After six more years of work he presented his final paper on this subject (Shaw & Davy, 1923a, 1923b). With the assistance of Norman Davy (1893-1973), the apparatus was improved.[7] The supports and suspending systems of the masses were changed. The experiment was carried on, and the observed effect was now very small: about 0.2×10^{-5} / °C, or possibly zero. Shaw's final conclusion was that, for the studied temperature range (from 0 °C to 250 °C) there is no temperature effect and G is constant.

No other papers were published on this subject, after this one. From this time onwards, it was generally agreed that gravitation does not depend on temperature – or, at least, that no large effect exists.

---

[7] I was unable to consult Norman Davy's MSc dissertation on this subject: *The effect of temperature on gravitative attraction.* University of London, 1929. It probably presented new improvements and results, but no paper was published afterwards containing those final researches on the subject.

8. CONCLUSION

At first sight, this long and arduous search led to nothing. But this is not strictly true. A null result is not equivalent to no result. Even before Shaw's experiments, several scientists supposed that the measured effect was spurious; nevertheless, Lindemann and Burton commented:

> In conclusion, we should like to express our admiration for Dr. Shaw's experimental work. We feel that as the result of such an elaborate research a null result is quite as important as, if less sensational than, a positive one. To have reduced - 3the apparent temperature coefficient of gravity from the 10 deduced from Prof. Boys' measurement to 1/80 of that value is certainly no mean achievement (Lindemann & Burton, 1917).

It is possible to list several methodological rules that were fulfilled by those researches and that show their scientific value:

a) Once someone (or several persons) belonging to the scientific community strongly suggest the possible existence of a measurable effect, it is desirable to search for such an effect. If it is found, good: a new physical phenomenon was discovered. If it is not found, this is also a valuable result, as it will bind speculative theory building and increase our knowledge about some constancies or conservations in nature.

b) Even if no theory suggests some correlation or effect, it is desirable to look for them by increasing the sensitivity of measurement procedures, by varying the conditions and bu producing greater changes of the independent parameters. The results of this search are of the same kind as in the former case.

c) In any circumstances, it is also desirable to develop new experimental techniques that allow the study of new situations, new effects, new materials, with increased accuracy and sensitivity, and so on. It is also desirable to suggest, analyze and test possible systematic errors in previously developed techniques, trying to measure, compensate and eliminate such errors, if they exist.

All those *desiderata* were fulfilled in the search for a temperature influence upon gravitation. They were certainly useful, increasing both our knowledge about nature and by the development of new techniques. One might criticize those researches by pointing out that too much time and money was lost and no effect was found. But in any relevant scientific research (an specially in fundamental research) it is not possible to know beforehand what the result will be like. Had Hicks' and Shaw's first results been confirmed, this would have been a very important discovery.

It is also relevant to remark that, after the general acceptance of General Relativity, there were strong theoretical grounds for stating that no large temperature influence on gravitation would exist. Indeed, since the source of gravitational effects, within General Relativity, is the stress-energy tensor, temperature may affect gravitational forces, since by varying temperature the energy of the field producing body may change. The effect will, of course, depend on the substance and on several other factors. For water, the effect should be about $4.6 \times 10^{-14}/°C$ – that is, eight orders of magnitude below the sensitivity of Shaw's experiments. It is very likely therefore that the acceptance of General Relativity in the 1920's (specially after the eclipse light-deflection test and the agreement between theoretical and empirical astronomical red-shifts) was a decisive reason for the disappearance of this kind of experimental research.

ACKNOWLEDGMENTS

The author is grateful to the Research Foundation of the São Paulo State (FAPESP) for a travel grant (1995-1996) that allowed him to have access to the material needed to complete this work. The author is also grateful to the Brazilian National Council of Scientific and Technological Development (CNPq) for supporting this research.

BIBLIOGRAPHIC REFERENCES

BAILY, Francis. Experiments with the torsion rod for determining the mean density of the earth. *Memoirs of the Royal Astronomical Society,* **14**: 1-120, i-ccxlviii, 1843.

BARTON, Edwin H. Gravitation and temperature. *Nature,* **97**: 461-462, 1916.

BARTON, Edwin H. [untitled]. *Nature,* **99**: 45, 1917.

BENTHAM, Muriel A. Some seventeenth century views concerning the nature of heat and cold. *Annals of Science,* **3**: 431-450, 1937.

BERTHELOT, Marcelin. Sur les registres de laboratoire de Lavoisier. *Comptes Rendus de l'Académie des Sciences de Paris,* **135**: 549-551, 1902.

BONDI, Herman. Negative mass in general relativity. *Reviews of Modern Physics,* **29**: 423-428, 1957.

BOYS, Charles Vernon. On the Newtonian constant of gravitation. *Nature,* **50**: 330-334, 366-368, 417-419, 571, 1894.

CORNU, Alfred Marie & BAILLE, Jean-Baptiste. Détermination nouvelle de la constante de l'attraction et de la densité moyenne de la Terre. *Comptes Rendus Hebdomadaires des Séances de l'Académie des Sciences de Paris,* **76**: 954-980, 1873.

CROOKES, William. On the atomic weight of thallium. *Philosophical Transactions of the Royal Society of London,* **163**: 277-330, 1873.

CROOKES, William. On attraction and repulsion resulting from radiation. *Philosophical Transactions of the Royal Society of London,* **164** (ii): 501-527, 1874.

CROOKES, William. On repulsion resulting from radiation. *Philosophical Transactions of the Royal Society of London,* **165** (ii): 519-547, 1875; **166** (ii): 325-354, 355-376, 1876.

DRUDE, Paul. Ueber Fernewirkungen. *Annalen der Physik,* **62**: i-xlix, 1897.

EINSTEIN, Albert. Relativitätsprinzip und die aus demselben gezogenen Folgerungen. *Jahrbuch der Radioaktivität und Elektronik*, **4**: 411-462, 1907; **5**: 98-99, 1908.

FORDYCE, George. An account of some experiments on the loss of weight in bodies on being melted or heated. *Philosophical Transactions of the Royal Society of London*, **75**: 361-365, 1785.

FRESNEL, Augustin. Note sur la répulsion que les corps échouffés exercent les uns sur les autres à des distances sensibles. *Annales de Chimie et Physique,* **29**: 57-61, 107-108, 1825. Reproduced in vol. 2, pp. 667-672: *Oeuvres complètes d'Augustin Fresnel* (ed. by H. de Senarmont, É. Verdet and L. Fresnel). Paris: Imprimérie Impériale, 1868.

HASENÖHRL, Fritz. Zur Theorie der Strahlung in bewegten Körpern.*Annalen der Physik*, [4] **15**: 344-370, 1904; **16**: 589-592, 1905.

HERAPATH, John. *Mathematical physics or the mathematical principles of natural philosophy*. 2 vols. London: Whittaker, 1847.

HICKS, William Mitchinson. On some irregularities in the values of the mean density of the Earth, as determined by Baily. *Proceedings of the Cambridge Philosophical Society* **5**: 156-161, 1883-6.

[LARMOR], Sir Joseph. Gravitation and temperature. *Nature,* **97**: 321, 1916.

[LARMOR], Sir Joseph. Gravitation and temperature. *Nature,* **97**: 421, 1916.

[LARMOR], Sir Joseph. Thermodynamics and gravitation. *Nature,* **99**: 44-45,1917.

LAVOISIER, Antoine Laurent. Nouvelles réflexions sur l'augmentation de poids qu'acquiérent, en brûlant, le soufre et le phosphore, et sur la cause à laquelle on doit l'attribuer. *Mémoires de l'Académie des Sciences de Paris,* 1783, p. 416. Reproduced in: Vol. II, pp. 616-22, *Oeuvres de Lavoisier*. Paris: Imprimérie Impériale, 1862.

LAVOISIER, Antoine Laurent; LAPLACE, Pierre-Simon. Mémoire sur la chaleur. *Mémoires de l'Académie des Sciences de Paris*, 1780, pp. 355-408. Reproduced in Vol. II, pp. 283-333, in *Oeuvres de Lavoisier*. Paris: Imprimérie Impériale, 1862.

LINDEMANN, Frederick A. & BURTON, Charles Vandeleur. The temperature coefficient of gravity. *Nature,* **98**: 349, 1917.

LODGE, Oliver Joseph. Gravitation and thermodynamics. *Nature,* **99**: 104, 1917.

NEWTON, Isaac. *Mathematical principles of natural philosophy.* Ed. F. Cajori. Berkeley: University of California, 1962.

OPPENHEIM, Samuel. Kritik des Newtonschen Gravitationsgesetzes [1920]. Pp. 80-158, in: SCHWARZSCHILD, Karl; OPPENHEIM, Samuel; DYCK, Walter von. (eds.). *Encyclopädie der mathematischen Wissenschaften*, Band VI.2.B – *Astronomie*. Leipzig: Teubner, 1922-1934.

PARTINGTON, James Riddick; MCKIE, Douglas. Historical studies on the phlogiston theory. *Annals of Science*, **2**: 361-404, 1937; **3**: 1-58, 337-371, 1938; **4**: 113-149, 1939.

PLANCK, Max. Zur Dynamik bewegter Systeme. *Sitzungsberichte den Akademie der Wissenschaften zu Berlin* (1907): 542-570; reproduced in *Annalen der Physik*, [4] **26:** 1-40, 1908.

POINCARÉ, Henri. Les limites de la loi de Newton. *Bulletin Astronomique,* **17**: 121-269, 1953.

POUILLET, Claude Servais Mathias. Note historique sur divers phénomènes d'attraction, de répulsion et de déviation qui ont été attribués à des causes singulières et qui s'expliquent naturellement par l'action de certains courants d'air dont on n'avait pas soupçonné l'existence. *Comptes Rendus des Séances de l'Académie des Sciences de Paris,* **29**: 245-249, 1849.

POYNTING, John Henry. Recent studies in gravitation. *Nature,* **62**: 403-408, 1900.

POYNTING, John Henry & PHILLIPS, Percy. An experiment with the balance to find if change of temperature has any effect upon weight. *Proceedings of the Royal Society of London,* **A 76**: 445-457, 1905.

REICHENBÄCHER, Ernst. Träge, schwere and felderzeugende Masse. *Zeitschrift für Physik,* **15**: 276-279, 1923.

REYNOLDS, Osborne. On the forces caused by the communication of heat between a surface and a gas; and on a new photometer. *Philosophical Transactions of the Royal Society of London,* **166**: 725-735, 1876.

SCHAGRIN, Morton L. Early observations and calculations on light pressure. *American Journal of Physics,* **42**: 927-940, 1974.

SCHUSTER, Arthur. On the nature of the force producing the motion of a body exposed to rays of heat and light. *Philosophical Transactions of the Royal Society of London,* **166**: 715-724, 1876.

SHAW, Philip E. The masses of heavenly bodies and the newtonianconstant. *Nature,* **96**: 143-144, 1915.

SHAW, Philip E. The newtonian constant of gravitation as affected by temperature. *Philosophical Transactions of the Royal Society of London,* **A 216**: 349-392, 1916(a).

SHAW, Philip E. Gravitation and temperature. *Nature,* **97**: 400-401, 1916(b).

SHAW, Philip E. [untitled answer to Lindemann and Burton]. *Nature,* **98**: 350, 1917(a).

SHAW, Philip E. Gravitation and thermodynamics. *Nature,* **99**: 84-85, 1917(b).

SHAW, Philip E. Gravitation and thermodynamics. *Nature,* **99**: 165, 1917(c).

SHAW, Philip E.; HAYES, Cecil. A special test on the temperature effect of gravitation. *Proceedings of the Physical Society,* **29**: 163-169, 1917.

SHAW, Philip E.; DAVY, Norman. The effect of temperature on gravitative attraction. *Physical Review,* **21**: 680-691, 1923(a).

SHAW, Philip E.; DAVY, Norman. The effect of temperature on gravitative attraction. *Proceedings of the Royal Society of London*, **A 102**: 46-47, 1923(b).

SOUTHERNS, Leonard. Experimental investigation as to dependence of gravity on temperature. *Proceedings of the Royal Society of London*, **A 78:** 392-403, 1907.

SPRAT, Thomas. *The history of the Royal-Society of London, for the improving of natural knowledge.* London: Royal Society, 1667. Reprinted by: St. Louis: Washington University, 1959.

STERNECK, Robert von. Untersuchungen über den Zusammenhang der Schwere unter der Erdoberfläche mit der Temperatur. *Sitzungsberichte der mathematische-naturwissenschaftliche Classe der Kaiserlichen Akademie der Wissenschaften,* **108**: 697-766, 1899.

TAYLOR, William Bower. Kinetic theories of gravitation. *Report of the Board of Regents of the Smithsonian Institution,* 1876, 205-282.

THOMSON, Benjamin (Count Rumford). An inquiry concerning the weight ascribed to heat. *Philosophical Transactions of the Royal Society of London,* **89**: 179-94, 1799; *Philosophical Magazine,* **5**: 162-174, 1799; *Bibliotheca Britannica,* **13**: 217-238, 1799.

TODD, George W. Thermodynamics and gravitation: a suggestion. *Nature,* **99**: 5-6, 1917(a).

TODD, George W. |untitled|. *Nature,* **99**: 104-105, 1917(b).

TOZZETTI, Giovanni Targioni (ed.). *Atti e memorie inedite dell'Accademia del Cimento.* Firenze: Giuseppe Tosani, 1780.

ZENNECK, Jonathan Adolf Wilhelm. Gravitation [1901]. Pp. 25-67, in: SOMMERFELD, Arnold (ed.). *Encyclopädie der mathematischen Wissenschaften.* Band V.1 – *Physik.* Leipzig: Teubner, 1903-1921.

PHILOSOPHY IN THE PHYSICS LABORATORY: MEASUREMENT THEORY VERSUS OPERATIONALISM

Roberto de Andrade Martins

Abstract: This paper presents Hermann van Helmholtz' approach to measurement theory, and discusses the possible uses and implications of his views in science education. This approach opposes the "black box" attitude towards measurement (operationalism). It takes into account the *a priori* conditions that should be imposed upon measurement procedures to obtain results that conform to some basic mathematical properties of physical magnitudes. Instead of the discredited but still popular empiricist view that the scientist should observe nature without any preconceived ideas, Helmholtz' theory of physical magnitudes shows that in basic measurement procedures it is necessary to introduce theoretical considerations. At the same time, this does not introduce a vicious circle in experimental testing.

Keywords: measurement theory; systematic errors; operationalism; science education; Helmholtz, Hermann von

1. INTRODUCTION

Among physicist and other scientists, the phrase "measurement theory" is usually associated to operationalism, inductivism and the study of random errors. Measuring instruments and, in general, scientific instruments, are regarded as "black boxes" that produce readings when applied to physical systems. The measuring instrument "defines" a physical

MARTINS, Roberto de Andrade. *Studies in History and Philosophy of Science I*. Extrema: Quamcumque Editum, 2021.

magnitude. Therefore, when two different instruments (or methods) are applied to the same object, they can produce different results, because they are not measuring the same property.

Every physics student learns that measurement is one of the most important activities in the physics laboratory. In the educational laboratory, measuring apparatuses are used to obtain measurements; they are instruments, and therefore they are never the subject of study in the physics lab. Textbooks about measurement and laboratory techniques usually dedicate a few paragraphs to systematic errors. Only random errors deserve a detailed study.

In this context, it is not clear what meaning can be ascribed to questions such as: Is the measuring instrument working properly and measuring what it was intended to measure? If a physical magnitude is *defined* by its measuring method, it is impossible to criticise or improve the method. If distances are measured (and operationally defined) by the instruments that measure distance, it is meaningless to ask whether the ruler measures distances *correctly* or not.

Measurement devices are not, however, blindly made instruments. Also, it is not necessary to have a blind faith in their performance. It is possible to analyse and test measurement instruments; and there is a general theory of measurement instruments and a general theory of systematic errors that can be useful in real-life laboratory.

However, the required background is not operationalism, but a completely different approach. Before the advent of operationalism in the early 20th century, there was already a deeper analysis of measurement theory provided by Hermann von Helmholtz (1821-1894) in 1887. This approach takes into account the *a priori* conditions that should be imposed upon measurement procedures in order to obtain results that conform to some basic properties of physical magnitudes. Helmholtz' approach was followed and developed by Norman Robert Campbell (1880–1949) in 1920. In more recent times, one can find an excellent textbook on this subject written by a philosopher: Brian David Ellis (1968).

This paper will present a short account of Helmholtz' approach to measurement theory, and will discuss the possible uses and implications of this approach in physics education.[1]

2. HELMHOLTZ' CONTRIBUTION

Hermann von Helmholtz' famous analysis of measurement is contained in his essay *Zählen und Messen erkenntnis-theoretisch betrachet* (Helmholtz, 1887). That writing was part of Helmholtz' discussion of Kant's views on the relations between science and experience. In some former papers, Helmholtz had argued that the axioms of geometry are not propositions given *a priori*, but rather that they are to be established or refuted through experience. In a similar vein, in his 1887 essay he discussed the axioms of arithmetic and attempted to unravel their empirical content.

The first part of Helmholtz' essay is dedicated to the discussion of number, arithmetic and counting. In the second part of his essay, he discusses measurement.

Helmholtz presented the following traditional axioms of arithmetic[2]:

Axiom I. If each of two magnitudes is equal to a third, they are equal to each other.
Axiom II. Associative law of addition: $(a+b)+c=a+(b+c)$
Axiom III. Commutative law of addition: $a+b=b+a$
Axiom IV. Equals added to equals give equals.

---

[1] This paper was presented at the Eighth International History, Philosophy, Sociology & Science Teaching Conference, University of Leeds, England, July 15 to 18, 2005. It is published here for the first time. I had previously published two papers that addressed some facets of the subject (Martins, 1984a; Martins, 1984b).

[2] Helmholtz proposed an additional axiom (Axiom VI): If two numbers are different, one of them must be higher than the other.

Axiom V. Equals added to unequals give unequals.

A large part of Helmholtz' paper is devoted to the discussion of those axioms and the mathematical concept of number. He shows, for instance – following Hermann and Robert Grassmann – that it is possible to reduce axioms II and III to another one, namely, $(a+b)+1=a+(b+1)$. This part of his essay will not be discussed here. Let it be said, however, that his approach to the foundation of arithmetic was regarded as naïve by most mathematicians of that time, and that this field of investigation soon gained rigour and clearness. Dedekind, Cantor and Frege despised Helmholtz' approach that mixed up mathematics with empirical issues (Darrigol, 2003, pp. 518, 557-561).

In the second part of his work, Helmholtz introduced the concept of *magnitude*: "Objects or attributes of objects, which, when compared with similar ones, permit the distinction of greater, equal, or smaller, we call magnitudes" (Helmholtz, 1930, p. 17). In some cases (but not always) it is possible to ascribe numbers to magnitudes. "The procedure whereby we find the denominate number we call the measuring of magnitudes" (Helmholtz, 1930, p. 17).

Under what conditions can numbers and their operations be applied to the relations of real objects and their magnitudes? He reduced this problem to two simpler ones: the empirical meaning of *equality* and of *addition* of magnitudes.

> 1. What is the objective meaning of declaring two objects in a certain relation equal?
> 2. What character must the physical combination of two objets have in order that we may consider comparable attributes of the same as united additively and these attributes accordingly as magnitudes which can be designated by denominate[3] numbers? (Helmholtz, 1930, p. 4)

---

[3] "Denominate" numbers are those accompanied by units.

2.1 The comparison of two magnitudes

The first point Helmholtz elucidates is the meaning of equality (or inequality) as applied to the attributes of objects. If we apply to this concept (physical equality) the same properties that he had already described in the case of numbers, then the new concept must obey the following properties:

(P1) if $a=b$ then $b=a$ (symmetry)

(P2) if $a=c$ and $b=c$ then $a=b$ (transitivity)

Now, suppose we have a method for comparing some attribute of different objects (for instance, their size or weight). If we compare objects A and B and conclude that some of their attributes (size, weight, etc.) are equal, we should expect that the comparison of B and A will also lead to the same conclusion, by (P1).

Suppose, for instance, that we compare the weights of two bodies using the kind of balance that was employed in the 19th century. If the two bodies A and B are in equilibrium, we conclude that their weights are equal. If we exchange the positions of A and B, the balance must remain in equilibrium. If this does not occur, the instrument is inadequate (Helmholtz, 1930, pp. 19-20).

Helmholtz pointed out that the second property (P2) does also have an empirical meaning (Helmholtz, 1930, p. 20). If we use a two-pan balance to compare the weights of A and C and we observe that they equilibrate each other; and if B and C do also equilibrate each other; we should expect that A and B will also equilibrate each other. If that does not occur, the instrument is inadequate.

Therefore, properties (P1) and (P2) can be used to test comparison instruments. A "correct" instrument should pass tests grounded upon properties (P1) and (P2). Of course, it is not possible to prove that the instrument is correct, but it may be possible to find out that it is incorrect. Those properties determine what physical relations we can recognise as relations of equality (Helmholtz, 1930, p. 22).

When a given comparison procedure is proposed and tested, the results can conform to the expected results (taking into account the mathematical properties of equality) or they may fail to conform to the predicted results. In the second case, the comparison procedure is rejected. However, in principle, the *mathematical properties of equality* might as well be rejected. Why they are never put to test? Because we have a preconceived (*a priori*) idea of mathematical equality which does not come from experience and that will not be invalidated by any observation.

Notice that Helmholtz introduced *comparison methods* in a way that precedes and is independent of ascertaining the *value* of the corresponding magnitudes. In our time, if we ask someone who is not familiar with measurement theory how could he/she determine whether two objects have the same length or weight, the most likely answer would be that we should measure both objects and compare their measurements. However, there are procedures which are prior to measurement proper and that allow us to *compare* objects as regards some specific quantities. We can see whether two persons are equally tall without measuring their sizes, by putting them side by side. We can test whether two bodies have the same weight by putting them on the two pans of an equal-arm balance, and checking whether the balance remains in equilibrium or not. This is a fundamental idea, that should be stressed in elementary physics courses.

2.2 Physical addition of two magnitudes

Mere comparison of two magnitudes can show whether they are equal or unequal, but does not provide measurements (numbers) to those magnitudes. If we are to describe those magnitudes by numbers, and if those numbers are to be used in arithmetical operations, they must obey some conditions. As the basis of all arithmetical operations is addition, Helmholtz first analysed "under what conditions we can express a physical combination of magnitudes of the same denomination as an addition" (Helmholtz, 1930, p. 22).

We add the lengths of two bodies by placing one of their ends in contact with each other and by putting the contact point and their ends in a straight line (Helmholtz, 1930, p. 23). In the case of weights, the physical combination of A and B is a mere juxtaposition of the two objects (Helmholtz, 1930, p. 22). It does not matter if the two objects are put side by side, or one on the top of the other, or slightly separated. If we put the two objects on one pan of a two-pan balance, they will always be equilibrated by the same object C on the other pan. Therefore, the physical combination of two objects as regards their weights does obey a law similar to the commutative law of addition (P3): $a+b=b+a$.

This is, of course, an empirical finding. If A on the top of B equilibrated C, but B on the top of A did not equilibrate C, then the physical combination of A and B as regards their weight would not be a mere juxtaposition of the two objects. That is: a given rule of physical combination of magnitudes can be tested by comparing it to the properties of numbers. If the rule does not pass the test, it is an inadequate rule and should be rejected.

For other properties, the physical combination that produces the addition of their magnitudes is different.

> We add resistances when we unite the wires one after the other so that the electricity conducted through them must flow through each successively. We add conductivity of the wires when we put the wires side by side and unite all their beginnings and also all their ends. (Helmholtz, 1930, p. 25)

It is possible to apply several tests, using other laws of arithmetical addition, such as the axiom "equals added to equals give equals" (P4): if $a=b$ and $c=d$ then $a+c=b+d$. That is, if bodies A and B are equal as regards their weight, and C and D do also equilibrate, then we should expect that the combination of $A+C$ will equilibrate $B+D$. Also, if $A+C$ equilibrates another body E, then we should expect that $A+D$, $B+C$ and $B+D$ should also equilibrate E. If that does not occur, then either the comparison rule (the balance) or the combination rule (the

143

juxtaposition of objects) is inadequate. If all those properties are obeyed, then the combination rule can be (temporarily) accepted as an adequate physical counterpart of arithmetical addition.

> A physical method of combining magnitudes of like kind can then be regarded as addition, if the result of the combination, compared as magnitudes of the same kind, is not changed, either by the interchange of single elements or by the exchange of members of the combinations with equal magnitudes of the same kind. (Helmholtz, 1930, p. 24)

It is not possible to apply this analysis to some other magnitudes, such as temperature and density, because when we join two objects with temperatures (or densities) A and B, the temperature (density) of the system does not become $C=A+B$.

Notice that we can physically add many (but not all) physical magnitudes. It is possible to add the volumes of two non-reacting liquids by putting them in the same vessel; it is possible to add the resistances of two pieces of wire joining them in series; it is possible to add potential differences in the same way. It is possible to add currents by joining parallel wires. It is possible to add the intensities of two incoherent radiation sources by their simultaneous action upon the same surface. Using those properties, it is possible to test the linearity of instruments that measure volume, resistance, potential difference, electric current, radiation intensity, etc.

Remark that there is an empirical component and an *a priori* component in the process of physical addition. According to our *concept* of physical magnitudes, we have need of associating numbers to magnitudes and to perform arithmetical operations with them. This would be pointless if there was no relationship between arithmetical operations and physical properties. It is meaningless, for instance, to perform arithmetical operations with telephone numbers. It is meaningful, however, to perform arithmetical operations with length measurements.

In the case of several physical magnitudes, it is possible to find physical operations that obey the same properties as arithmetical addition. For instance: if we join two objects together, the weight of the new (compound) object is the sum of the masses of the two objects. This property was not derived from measurement: it is an *a priori* requirement imposed upon measurement. Any balance (even the most expensive model of a beautifully illustrated catalogue) that refuses to obey this law of addition of weight should be rejected as inexact. However, it is an empirical matter to check whether this addition rule does really obey the same rules as the arithmetical addition.

If we know how to compare two magnitudes of the same kind without measuring them, and if we know how to add two magnitudes of the same kind by joining two physical systems in a specific way, then we can create a measurement procedure for that magnitude. The typical instance is again provided by the old equal-arms balance together with a set of standard weights. It is possible to combine the standard weights and it is possible to compare their combination to any given body. In that way, it is possible to ascribe numbers (measurements) to those bodies.

3. PRECEDENTS OF HELMHOLTZ' WORK

Olivier Darrigol published a detailed analysis of the authors who probably influenced Helmholtz' theory of measurement, and the influence exerted by Helmholtz' ideas (Darrigol, 2003).

Helmholtz was inspired by mathematicians such as the Grassmann brothers, Hermann and Robert and Paul Du Bois-Reymond and by physiologists / psychologists who discussed the possibility of measuring psychological quantities (Darrigol, 2003, pp. 520-541). He was probably influenced by Ernst Mach and James Clerk Maxwell, too. There are several similarities between Maxwell's previous discussion of measurement of electric charge and temperature and Helmholtz' ideas (Darrigol, 2003, pp. 541, 548).

In his *Theory of heat* (1871) Maxwell discussed the measurement of temperature and emphasized that the equality between the temperatures of two bodies is a concept more fundamental than the value of their temperatures, and that it can be ascertained by putting the two bodies in contact and checking whether there was any heat flow from one to the other. However, if this test is to be used for ascertaining equality of temperature, it must obey a testable law: if the temperatures of A and B are equal to the temperature of C, then the temperatures of A and B should also be equal. If a piece of iron is plunge in a vessel of water and it is noticed that they are in thermal equilibrium; and if the same piece of iron, being transferred to a vessel full of oil, is observed to be in thermal equilibrium with the oil; than it is possible to predict (and to check) that the water and the oil will be in thermal equilibrium (Darrigol, 2003, p. 542). Nowadays this law is sometimes called "the zeroth law of thermodynamics". It is seldom ascribed to Maxwell, and its relation to the general theory of measurement is never mentioned.

Maxwell also remarked that it is impossible to add two bodies at temperatures P and Q producing a third body at the temperature $P+Q$. If there were such a procedure, then it would be possible to measure temperatures in the same way we measure mass or length.

Although there were some precedents to the ideas presented by Helmholtz, as was shown above, it seems that his essay was the first systematic discussion of the measurability of physical properties (Darrigol, 2003, p. 516).

4. CAMPBELL'S CONTRIBUTION

Helmholtz' analysis is very useful to check basic measurement procedures in order to detect systematic errors. In his paper, Helmholtz did not discuss random errors. Also, he did not analyse other kinds of measurement, that apply to magnitudes for which there is not a procedure of physical addition (such as density). Norman Campbell, in his 1920 book *Physics, the elements*, presented a more detailed (and, in some senses, more satisfactory) account of measurement. In his book, Campbell never referred to Helmholtz; however, directly or indirectly, he certainly suffered his influence.

Campbell introduced the name "fundamental measurement" to characterize the direct measurement procedures (those that do not depend on the measurement of other quantities), which are grounded upon comparison and physical addition (Campbell, 1957, pp. 277-278) – that is, those that were studied by Helmholtz. Of course, Helmholtz knew (and stated) that some quantities (such as temperature and density) cannot be measured in the same way as weight or length, because there is no known way of producing the physical addition of two systems as regards those properties. However, Helmholtz did not give *names* do the different cases, nor did he elucidate other forms of measurement. Campbell distinguished several cases, and gave the name of *derived measurement* to the measurement of quantities such as density.

Instead of a relation of *equality* between two quantities, Campbell used another comparison relation, that of *greater than*, or *smaller than*. The physical order relation must be a transitive and asymmetrical relation, as the corresponding mathematical relation – and it is possible to *test* whether a given physical comparison method does obey or does not obey those properties (Campbell, 1957, pp. 270-274).

Helmholtz did not discuss random errors and the difficulty they introduce in fundamental measurement. It may happen that some comparison method shows that A and B are equal (relative to some magnitude), and that B and C are also equal, but A and C are different. In order do deal with this situation it is necessary to introduce the concept of errors of measurement (or uncertainty). The comparison between two objects can only lead to the result that their difference is *smaller than the error of measurement*, but cannot establish that they are *equal*. Campbell developed the analysis of measurement taking into account the existence of errors (Campbell, 1957, chapter 16), while Helmholtz did not.

Although there are relevant differences between Campbell's and Helmholtz' approaches, both authors emphasise that the fundamental measurement of physical magnitudes involves operations that must obey a set of laws isomorphic to those of arithmetic[4]; and that it is possible to check whether a given operation (or instrument) does obey such a law, so that systematic errors (or "methodical errors", according to Campbell) can be eventually found.

If we take into account random errors, Helmholtz rules must be corrected. Suppose we measure the weight of two objects using a balance, and that their measurements are repeated several times, and their weights are $A \pm a$ and $B \pm b$. When both objects are put together on the balance pan, we should obtain a weight C compatible with $(A+B) \pm (a^2+b^2)^{1/2}$. Otherwise, we should conclude that the balance has a systematic error, since it is unable to add.

In the same way, suppose we measure the thickness of two plates using a micrometer, and obtain the values $A \pm a$ and $B \pm b$. When both plates are put together and measured with the same micrometer, we should obtain a thickness $C \pm c$ compatible with $(A+B) \pm (a^2+b^2)^{1/2}$.

It would be possible to present many other relevant features of Helmholtz' and Campbell's theories of measurement. Those that were shown here are sufficient, however, to exhibit the central ideas of this approach to measurement theory.

5. THE EDUCATIONAL USE OF MEASUREMENT THEORY

The discussion of these and other fundamental issues concerning the theory of physical magnitudes and their measurement can be introduced in our teaching practice and

---

[4] As a matter of fact, there is not an *isomorphism* between the physical quantities of the objects and arithmetic, but a weaker kind of *morphism*. Those distinctions will not be introduced here, however. They are discussed in several works in the mathematical tradition of measurement theory, beginning with Patrick Suppes (1951). A nice historical analysis of measurement theory which discusses those issues can be found in José Díez (1997).

brings a new light to experimental science. Instead of the discredited but still popular empiricist view that the scientist should observe nature without any preconceived ideas, the theory of measurement shows that it is necessary to introduce theoretical considerations in basic measurement procedures. In addition to random errors there are other kinds of errors, and there is a *theory* about the way instruments should behave that allow us to test them and to impose some conditions before we adopt a measuring instrument.

It is necessary to remark that the idea that there is a theory underlying our measuring instruments is indeed generally recognized nowadays. This truism is part of the post-modern philosophy of science and is usually presented as an argument against the objectivity of quantitative science. If theories are needed to build measurement instruments, then the measurements obtained with the use of those instruments are influenced by theory, and cannot be used to *test* a theory, because this would entail a vicious circle: the theory should be accepted because it was confirmed by some measurements, and the measurements should be accepted because they are grounded upon the theory.

There is no vicious circle, however, because the theory under the measurement apparatus is not the same theory that is assessed with the use of that apparatus. There are testable physical laws underlying the functioning of the measurement apparatus, but those physical laws are the laws of measurement, analogous to the laws of arithmetic, such as those presented by Helmholtz. Taking into account the theory of measurement, the devices can be tested (i.e., checked for systematic errors) before being applied to the test of scientific theories.

An acquaintance with measurement theory can provide a more adequate view of the nature of experimental science, and it can also provide an effective help in discussing and searching for systematic errors in the physics laboratory.

Both Helmholtz and Campbell believed that the theory of measurement could be useful in physics courses. Helmholtz reproduced his analysis of measurement in two of his textbooks, and Campbell explicitly recommended the inclusion of a general discussion of measurement in introductory physics courses (Darrigol, 2003, pp. 554-555, 569-570). However, neither Helmholtz nor Campbell was successful in disseminating those ideas among physicists and physics teachers.

The theory of measurement developed by Helmholtz, Campbell and other authors was not integrated into physics textbooks. It was discussed by philosophers and by psychologists and incorporated in their works. In the context of physics teaching, either measurement theory is completely ignored, or it appears under the form of *operationalism*.

6. THE OPERATIONAL APPROACH IN PHYSICS TEACHING

The operational approach to measurement emerged in the early 20th century from the work of physicists, and had a strong influence upon other fields. The main representative of this approach was Percy W. Bridgman, who published a book (*The logic of modern physics*) and several papers on his theory.

According to operationalism, the measuring procedure *defines* a scientific magnitude. If we do not know how to measure something, it should not be regarded as a scientific magnitude. If we know how to measure it, then we know all that can be known about that magnitude. The measurement procedure, being a definition, is a *convention*. We can choose whatever procedure we want.

Of course, there *was* a positive characteristic of operationalism: it emphasized the importance and utility of describing how is it possible to measure a given magnitude – when that is possible. There are many relevant physical magnitudes (such as the vector potential of electromagnetism) that cannot be measured at all. If we accepted the operational point of view, we should reject them. Now, it would be very useful to have some procedure to measure those quantities;

however, even though we cannot measure them, they are useful in physics.

Even in the cases when we *can* measure a magnitude, the measurement procedure is not a *definition* of that quantity, and it is not arbitrary. If the instrument *defined* the magnitude, it would be impossible to say that the instrument is *wrong*. That is the relevant point that is stressed in the present paper. There exists a *measurement theory*, and that theory can tell use whether a given measurement procedure is acceptable or not (at least in some cases).

Up to the decade of 1950, operationalism was popular among both scientists and philosophers. During the second half of the 20th century, however, it suffered heavy attacks from philosophers and was completely discredited.

Although rejected by philosophers, operationalism retained its appeal to physicists. In the last decades of the 20th century operationalism still appeared explicitly in educational articles as the accepted theory of measurement. In a paper describing some novelties introduced in the physics laboratory at the University of California, Berkeley, in 1979, the authors described that they expected the students to learn some general skills, "those which practicing scientists commonly use, but which most students do not possess". The "general skills" were:

> (i) **Being able to use operational definitions to relate symbolic concepts to observable quantities**. This skill subsumes the ability to estimate or measure important physical quantities at various levels of precision. (ii) Being able to estimate the errors of quantities obtained from measurements. This skill involves applying habitually some qualitative or semiquantitative statistical notions, without any resort to excessive mathematical formalism. (iii) Knowing and applying some generally useful measuring techniques for improving reliability and precision, for example, such techniques include making repeated measurements, using independent measurement methods, or applying comparison

or "null" methods. (Reif & St. John, 1979, p. 950; my emphasis).

It is possible to find more recent papers published in physical journals that still regard operationalism as a viable theory of measurement (Delaney, 1999), although physicists have been told by Mario Bunge (and other authors) that operationalism is scientifically and philosophically inadequate (Bunge, 1967). Several generations of physicists and physics teachers, all over the world, have studied Resnick and Halliday's physics textbook (Resnick & Halliday, 1966). Both in the first and in the following editions, the first chapter of that treatise deals with measurement. In the older edition, the approach used by the authors is operationalism:

> For the purposes of physics, the basic quantities must be defined clearly and precisely. One view is that the definition of a physical quantity has been given when the procedures for measuring that quantity have been given. This is the *operational point of view* because the definition is, at root, a set of laboratory operations leading to a number with a unit. The operations may include mathematical calculations.
>
> [...]
>
> Examples of quantities usually viewed as fundamental are length and time. Their operational definitions involve two steps: first, the choice of a *standard*, and second, the establishment of procedures for comparing the standard to the quantity to be measured so that a number and a unit are determined as the measure of that quantity. (Resnick & Halliday, 1966, vol. 1, p. 2)

In more recent editions of this textbook, no explicit reference is made to the *operational point of view*, but the shallow concept of measurement presented in that manual is the same: the most important step in measurement is choosing a standard, and then, in some way that is not elucidated, the measured system is compared to the standard and there results some number. Notice that this was the way physics textbooks introduced measurement, before Helmholtz; and this is the way measurement is still presented by most physics teachers (Darrigol, 2003, p. 565).

The common attitude transmitted by the best physics laboratory textbooks is to ignore altogether the existence of measurement theory, paying attention mostly to statistics, and mentioning systematic errors *en passant*.

> Random errors may be estimated by statistical methods, which are discussed in the next two chapters. Systematic errors do not lend themselves to any such clear-cut treatment. Your safest course is to regard them as effects to be discovered and eliminated. There is no general rule for doing this. It is a case of thinking about the particular method of doing an experiment and of always being suspicious of the apparatus. We shall try to point out common sources of systematic error in this book, but in this matter there is no substitute for experience. (Squires, 1991, p. 11)

As a matter of fact, *there are general rules* that can be used to search for systematic errors. It is true that no rule will detect *all* systematic errors, but *many* systematic errors can be found if one follows some simple rules of measurement theory.

A recent multi-author paper that appeared recently in *The Physics Teacher* addresses the teaching of measurement (Allie et al., 2003). The main concern of the paper is how to teach students to deal with uncertainty in the introductory physics laboratory. Although the article does mention systematic errors, the approach of the authors is probabilistic and they do not address measurement theory.

Nowadays, measurement theory is ignored in scientific education. It is not mentioned in *Science for all Americans / Project 2061* of the American Association for the Advancement of Science. According to the guidelines of that project, it is expected that everyone should acquire the ability to "use appropriate instrument to make direct measurements of length, volume, weight, time interval, and temperature [...]" and also to "take readings from standard meter displays, both analog and digital, and make prescribed settings on dials, meters, and switches" (Rutherford, 1990, pp. 191-192). However, *understanding* the instruments and the principles underlying measurement is not included in the aims of that Project.

In the context of science education, few warnings have been published regarding the problems of operationalism. One nice exception can be found in a paper by Jorge Paruelo, where the author calls the attention to contradictions that can be found in physics textbooks which adopt the operational point of view:

> The contradiction is a consequence of not distinguishing the operationalization of a magnitude – that is, defining an operative mechanism to detect the presence of the magnitude, or to measure its quantity – and operationalism – that is, defining the magnitude by the aforementioned operation. A correct epistemological analysis of this subject will allow the development of a new route for teaching what a physical magnitude is and how it is detected. (Paruelo, 2003, p. 331)

I hope that the present paper will help calling the attention of physicists and educators to measurement theory, and will assist in improving the teaching of experimental physics. I do not claim that the ideas presented here can solve all the educational problems concerning the science laboratory. I do claim, however, that the subject of measurement is almost exclusively viewed by teachers and students as the blind use of measurement instruments and later statistical manipulation of data and that this is an inadequate view. Of course, even on that viewpoint there are educational problems – for instance, students have difficulty in understanding the need for repeating measurements and in dealing and interpreting uncertainties, and it is useful to investigate how those difficulties can be circumvented (Rollnick *et al.*, 2002). However, the current approach, even when it is successful, leads to an incomplete understanding of measurement. The study of measurement theory (in the sense offered in this paper) and the search for systematic errors that can be guided by that theory may contribute to broaden the understanding of physical measurement and to inspire a more adequate education of future scientists.

7. RESOURCES FOR STUDYING MEASUREMENT THEORY

This paper presented some general information concerning Helmholtz' theory of measurement and attempted to motivate further study and use of that approach. Any person interested in studying and applying this theory in physics education will need additional literature on the subject. What can I suggest?

Helmholtz' work was a landmark, but it is not the best presentation of his approach. It can be used, but one should keep in mind its several limitations, such as the ones that were described above.

Norman Campbell's presentation of measurement theory is remarkably clear, well organized, and written in an accessible style. It is more complete than Helmholtz' original essay and I do warmly recommend it. Although Campbell's *Foundations of science* was originally published almost one century ago, it is still very useful as an introduction to measurement theory, for physicists. It is available as a reprint (Campbell, 1957).

Brian Ellis, a philosopher, published in 1968 a very useful book on *Basic concepts of measurement* (Ellis, 1968). His work was not as rigorous as later developments in measurement theory, but in some sense that is an advantage: many recent books and papers on measurement theory are hermetic and can be described as scientifically sterile. They would not appeal to a scientist wanting to improve his scientific research or his teaching. Ellis' book is well written and, although the author was a philosopher, he did present the main ideas in such a way that physicists and other scientists will have no difficulty in understanding and enjoying most of the book.

I would not recommend to physicists and physics teacher the three volumes of *The foundations of measurement*, the current "Bible" of measurement theory (Krantz, Luce, Suppes &

Tversky, 1971-1990). This fundamental treatise is a very nice contribution to the current view of measurement, but its style does not seem suitable for most working scientists. Much of current publications on measurement theory only discuss the mathematical side (the axioms that the physical operations should mirror) or philosophical issues, but do not address the central point discussed here – the possibility of *testing* a measurement procedure by imposing that it should obey some rules analogous to those of arithmetic operations.

There are many different issues related to measurement theory. Some of them (such as those discussed in the present paper) can be perceived as directly relevant to the scientific practice, having important consequences for the practice of experimental physics. Other issues, although they have a strong appeal to measurement theory experts, are more remotely associated to the practice of experimental science. From the philosophical point of view, there are relevant *ontological issues*: Do physical objects *have* quantitative properties, or are quantities something extrinsic to physical objects? Some papers present a nice account of several epistemological problems of measurement – most of them having little relation to the scientific experimental practice (Mari, 2003; Boniolo, 2002). Therefore, the relevance of a large part of that literature to the practice of the physics laboratory is not very high.

8. CONCLUDING COMMENTS

This paper maintains that a specific part of philosophy of science can be improve the understanding and practice of measurement and that for that reason it would be valuable to introduce it in science education.

Sometimes the use of philosophy of science in education is viewed as a mere transmission of the views adopted by some specific philosophers, with no regard to the specific needs of science education. I agree that philosophy is useful and should be taught, but sometimes it may fail improving science education. Borrowing a phrase coined by Jorge Paruelo, what is urgently needed is an *applied philosophy of science* (PARUELO, 2003, p. 334), selecting and adapting the philosophical knowledge (and techniques) to science teaching. The teaching of the elements of measurement theory in the physics laboratory may be a fine step in this direction. Helmholtz' approach to measurement can be easily taught in undergraduate physics courses. We introduced it successfully in Brazil, several years ago. This theory of measurement provides a more adequate basis for the physical laboratory than operationalism and may improve the students' understanding of experimental physics.

ACKNOWLEDGEMENTS

The author is grateful to the Brazilian National Council for Scientific and Technological Development (CNPq) and to the São Paulo State Research Foundation (FAPESP) for supporting this research.

BIBLIOGRAPHIC REFERENCES

ALLIE, Saalih; BUFFLER, Andy; Campbell, Bob; LUBBEN, Fred; EVANGELINOS, Dimitris; PSILLOS, Dimitris; VALASSIADES, Odysseas. Teaching measurement in the introductory physics laboratory. *The Physics Teacher* **41**: 23-30, October 2003.

BONIOLO, Giovanni. On properties and processes of numerical assignment: an empiricist approach. *Axiomatics* **13**: 147-162, 2002.

BRIDGMAN, Percy W. *The logic of modern physics*. London: MacMillan, 1927.

BUNGE, Mario. Physical axiomatics. *Reviews of Modern Physics* **39** (2) 463-474, 1967.

CAMPBELL, Norman R. *Foundations of science: the philosophy of theory and experiment*. New York: Dover, 1957. [a reprint of *Physics, the elements*. Cambridge, Cambridge University Press, 1920]

DARRIGOL, Olivier. Number and measurement: Hermann von Helmholtz at the crossroads of mathematics, physics, and psychology. *Studies in the History and Philosophy of Science* **34**: 515-573, 2003.

DELANEY, William. Limitation of operational definitions. *International Journal of Theoretical Physics* **38** (6): 1757-1762, 1999.

DÍEZ, José A. A hundred years of numbers. An historical introduction to measurement theory 1887-1990. *Studies in the History and Philosophy of Science* **28** (1): 167-185; (2): 237-265, 1997.

ELLIS, Brian David. *Basic concepts of measurement.* Cambridge: Cambridge University Press, 1968.

HELMHOLTZ, Hermann von. *Counting and measuring.* Translated by C. L. Bryan; intr. and notes by H. T. Davies. New York: D. Van Nostrand, 1930.

KRANTZ, David H.; LUCE, Robert Duncan; SUPPES, Patrick; TVERSKY, Amos. *The foundations of measurement.* New York: Academic Press, 1971-1990. 3 vols.

MARI, Luca. Epistemology of measurement. *Measurement* **34**: 17-30, 2003.

MARTINS, Roberto de Andrade. Os elementos apriorísticos dos processos de medida. *Revista de Ensino de Física* **6** (2): 35-51, 1984(a).

MARTINS, Roberto de Andrade. Measurement and the mathematical role of scientific magnitudes. *Manuscrito* **7** (2): 71-84, 1984(b).

MICHELL, Joel. The origins of the representational theory of measurement: Helmholtz, Hölder, and Russell. *Studies in the History and Philosophy of Science* **24** (2): 185-206, 1993.

MICHELL, Joel. Bertrand Russell's 1897 critique of the traditional theory of measurement. *Synthese* **110**: 257-276, 1997.

MICHELL, Joel; ERNST, Catherine. The axioms of quantity and the theory of measurement. Translated from part I [and

II] of Otto Hölder's "Die Axiome der Quantität und die Lehre vom Mass". *Journal of Mathematical Psychology* **40**: 235-252, 1996; **41**: 345-356, 1997.

PARUELO, Jorge. Enseñanza de las ciencias y filosofía. *Enseñanza de las Ciencias* **21** (2): 329-335, 2003.

REIF, Frederick; ST. JOHN, Mark. Teaching physicists' thinking skills in the laboratory. *American Journal of Physics* **47** (11): 950-957, 1979.

RESNICK, Robert; HALLIDAY, David. *Physics*. New York: John Wiley, 1966.

ROLLNICK, Marissa; LUBBEN, Fred; LOTZ, Sandra; DLAMINI, Betty. What do underprepared students learn about measurement from introductory laboratory work? *Research in Science Education* **32**: 1-18, 2002.

RUTHERFORD, F. James (ed.). *Science for all Americans: Project 2061*. New York: Oxford University Press / American Association for the Advancement of Science, 1990.

SQUIRES, Gordon Leslie. *Practical physics*. 3rd edition. Cambridge: Cambridge University Press, 1991.

SUPPES, Patrick. A set of independent axioms for extensive quantities. *Portugaliae Mathematica* **10**: 163-172, 1951.